Radar Propagation at Low Altitudes

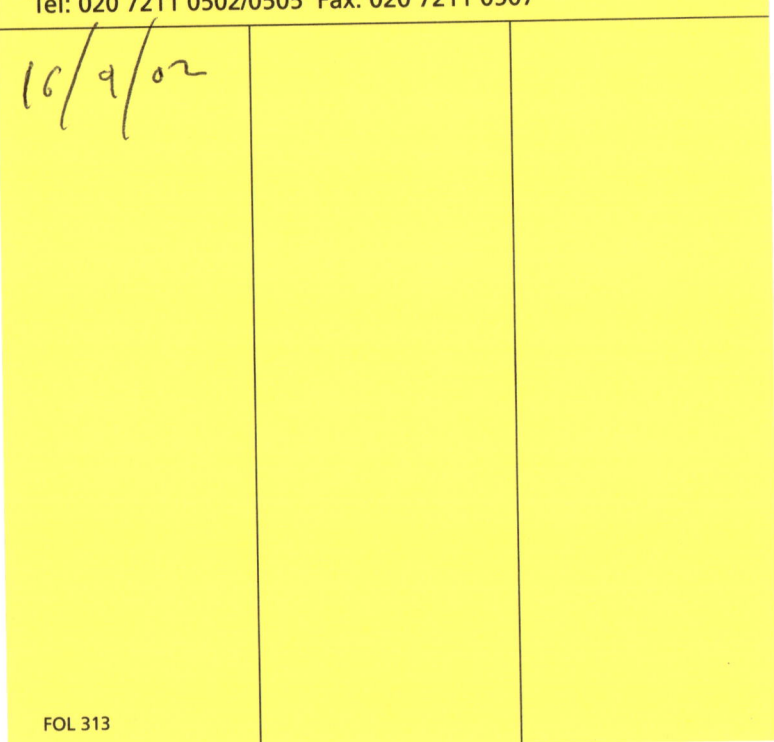

M.L. Meeks

Lincoln Laboratory
Massachusetts Institute
of Technology

Radar Propagation at Low Altitudes

artech

To Louise

ARTECH HOUSE, INC.
610 Washington St.
Dedham, MA

International Standard Book Number: 0-38006-118-1
Library of Congress Catalog Card Number: 82-72894

Table of Contents

Preface

Work on this monograph began during the summer of 1978 when I set out to learn what I could about electromagnetic-wave propagation near the surface of the earth. The first step in such a task, of course, is to review the literature and to consult people with first-hand experience in the field. Writing this monograph has helped to integrate my growing knowledge of the subject; the bibliography was compiled as I read through an expanding collection of journal articles and reports. A framework around which the information in these documents could be fitted was provided by selected reading in the books referenced in this monograph. Although the text mostly reviews the subject, two sections are essentially new: the analysis of the combined effects of reflection and diffraction in Section 6.3 and the numerical evaluation of the series solution for spherical-earth diffraction in Section 6.4.

During the later stages of writing, we began a series of field measurements of low-altitude propagation at Lincoln Laboratory. The experience gained making these measurements led to revisions and additions to the text that should make the treatment more useful. I have examined all the entries in the bibliography, which includes references published through 1981, but I cannot claim to thoroughly understand the contents of all of these. The reader will find it profitable to investigate the papers referenced for a subject of direct interest.

A number of people have helped me understand this subject. Among my collegues at Lincoln Laboratory, I would like to thank particularly John Delaney, James Evans, Ira Gilbert, Henry Helmken, Roger Reed, and John Ruze for valuable discussions. I am particularly grateful to David K. Barton of Raytheon Co. for initial orientation and many helpful suggestions as I pursued the subject. Jen A. Kong of Massachusetts Institute of Technology provided guidance to current literature on the reflection properties of soil, water, and snow. A. Vincent Mrstik of the General Research Corp. clarified questions concerning terrain reflection. Alfonso Malaga of Signatron, Inc. gave valuable insight into the problems of diffraction by a dielectric cylinder and diffraction by multiple knife-edges. Robert K. Crane of Dartmouth College provided information on attenuation produced by rain, clouds, and fog.

Many of the figures appearing in this monograph were generated by computer at Lincoln Laboratory, and I gratefully acknowledge the programming efforts of Mary Beth Carlson, Janet W. Hazel, and Gerald L. McCaffrey. The programs in Appendices A and B were written by Gerald McCaffrey.

For continued encouragement and administrative support throughout this work I thank Carl E. Nielsen, Jr. of Lincoln Laboratory. This work was supported in part by the Defense Advanced Research Projects Agency.

Introduction

This monograph is intended as a tutorial review of the physics of radio propagation at low altitudes over various kinds of terrain. The primary objective is to bring together an account of what is known about the propagation effects that determine the detection performance of ground-based radars against aircraft targets flying at low altitudes. However, the treatment here applies equally well to ground-based telecommunication problems.

The propagation of electromagnetic waves has been actively studied for over 50 years with a number of objectives. These include telecommunication and television coverage prediction, microwave-link design, studies of mobile radio communication, and radar performance and design studies. We have reviewed the literature in these fields, and we include in Appendix D a bibliography referencing, in separate sections, books, journal articles, and technical reports on this subject. Document references in the bibliography are listed alphabetically by the names of the first-listed author; journal articles and reports are additionally indexed by subject. References cited in the text distinguish between books, journal articles, and technical reports. For books the authors' names appear with the capital letter B before the year designation, for example Born and Wolf (B 1959); for journal articles the authors' names appear with capital J before the year designation, for example, Day and Trolese (J 1950). Finally, for technical reports the capital letter R appears before the year designation; as in, Longley and Rice (R 1968). Papers that have appeared in published proceedings of conferences have been considered as journal articles, but when conference proceedings are unpublished or not generally available, the papers have been classified as technical reports.

The physical phenomena that govern propagation — refraction, diffraction and reflection (multipath) — are treated in separate sections. The final section describes some simple propagation models that combine the fundamental physical phenomena, and Appendices A, B, and C present computer programs for these models.

Notwithstanding the work that has been done on VHF, UHF, and microwave propagation, there remain a number of fundamental, unanswered questions that are crucial to the complete understanding of low-altitude propagation. These questions concern the reflection properties of slightly rough surfaces at low grazing angles, including the reflection coefficient of various kinds of vegetative ground cover (see Chapter 4) and problems associated with multiple diffraction effects along a terrain profile (see Chapter 5).

2. THE PATTERN-PROPAGATION FACTOR

In this section we formulate the propagation problem associated with a ground-based radar searching for incoming aircraft at very low altitudes. We consider a radio frequency range from about 100 to 10,000 MHz (VHF through X-band) and assume a geometry in which the radar site has been selected for optimum coverage and the radar antenna has been mounted at a height of 30 m or less above local terrain. The aircraft, we assume, will fly at some altitude between 30 and 300 m above ground level. This search geometry clearly represents an extreme case of low-angle propagation.

The propagation effects produced by the earth's surface and atmosphere are taken into account by introducing the *pattern-propagation factor* F. As the name suggests, F takes into account the pattern of the transmitting antenna along with the propagation effects. The pattern-propagation factor is defined as the ratio of the electric field at the target to the electric field at the range of the target in free space and on the axis of the antenna beam. We can write this as follows:

$$F = \left| \frac{E}{E_o} \right| \tag{2.1}$$

where E_o is the magnitude of the free-space field at a given point when the antenna is pointed toward the point, and E is the field to be investigated at the point in question. In problems involving the propagation at low altitudes, it is frequently unnecessary to include the effects of antenna pattern in F because the angles involved are small compared with a beamwidth, and it would be more appropriate to call F simply the propagation factor in these cases. A detailed discussion of the pattern-propagation function is given by Kerr (B 1957), pp. 34-41, where the effect of antenna pattern is illustrated in the case of medium- and high-altitude targets.

The principle of reciprocity makes it unnecessary to distinguish between the pattern-propagation factor from radar to target and the factor from target back to radar. These factors are equal, and both may be represented by F. Thus the radar equation with propagation effects included may be written as:

$$P_r = \frac{P_t AG}{(4\pi)^2} \frac{\sigma}{R^4} F^4 \tag{2.2}$$

where we have designated the transmitted and received powers by P_t and P_r respectively, the effective aperture and gain of the radar antenna by A and G, the range by R, and the backscatter cross section of the target by σ.

In the case of one-way propagation over a range R, the received power is proportional to F^2 and R^{-2} rather than F^4 and R^{-4} in the two-way (radar) case. Reciprocity thus justifies applying the results of telecommunication studies to the problem of radar detection. In fact propagation measurements for the purpose of investigating radar performance are better made over one-way paths.

Referring to the radar equation, we can see that if propagation over terrain is compared with propagation in free space, the difference in return power from the target, measured in decibels, is given by $40 \times \log F$. Alternatively, if a minimum detectable return power $P_r(\min)$ is specified, then the *maximum detection range* along the axis of the radar beam in free space, R_{max} (free space), is given by

$$R_{max}\text{ (free space)} = \left(\frac{P_t AG}{(4\pi)^2} \frac{\sigma}{P_r(\min)} \right)^{1/4} \tag{2.3}$$

and for propagation over terrain, the maximum range becomes

$$R_{max} = F\, R_{max}\,(\text{free space}) \cdot \qquad (2.4)$$

Hence, F is the factor by which the maximum detection range in free space is altered by propagation effects.

3. REFRACTION EFFECTS

The *index of refraction* n of the atmosphere depends on the *pressure* P, the *temperature* T, and the *partial pressure of water vapor* C. Over the frequency range with which we are concerned, the index of refraction is essentially independent of frequency and given by

$$n = 1 + N \times 10^{-6} \qquad (3.1a)$$

where

$$N = 77.6\,(P/T) + 3.73 \times 10^{5}\,(e/T^{2}) \cdot \qquad (3.1b)$$

Here P and e are expressed in millibars and T in degrees Kelvin. The quantity N, the *radio refractivity*, is commonly referred to as being expressed in N units as given by Equation (3.1b). The quantities P and T can be measured directly, but the water-vapor content is usually measured indirectly by hygrometers (relative humidity), psychrometers (wet-bulb temperature), or dew-point devices (dew-point temperature). With any of these measurements, the *saturation vapor pressure* is required to convert the measured quantity to vapor pressure. Figure 3.1 shows the saturated vapor pressure of water plotted as a function of (T—273), the temperature on the Celsius scale. The contribution of the water-vapor component to the radio refractivity must be relatively small for cold air because the saturated vapor pressure is small as Figure 3.1 shows. But for warm air the relative humidity has a strong influence on the value of N.

Figure 3.2 shows the radio refractivity at sea level as a function of temperature for relative-humidity values 0, 30, 70 and 100%. We see that the water-vapor component of the refractivity increases exponentially with increasing temperature.

There is no corresponding humidity component in the atmospheric index of refraction at optical wavelengths.* For visible light the

*The electric dipole moment of water molecules can be reoriented by a radio-frequency electric field, but at optical frequencies the field reverses direction too rapidly for water molecules to follow.

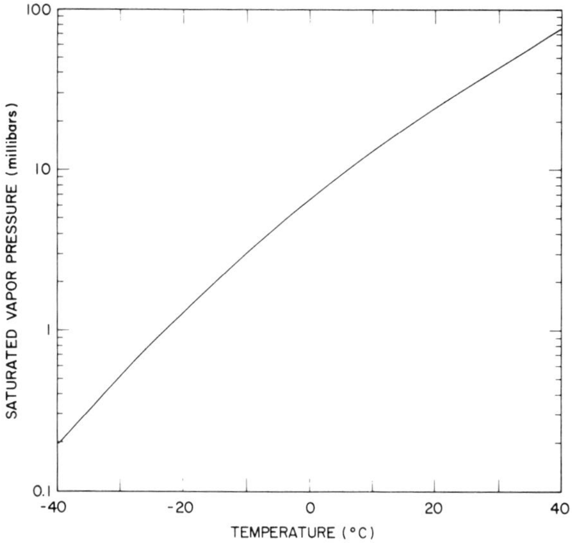

Fig. 3.1 The saturated vapor pressure of H_2O in millibars *vs.* temperature (Celsius). Over the temperature range from -40C° to +40° the pressure of saturated water vapor increases by nearly three orders of magnitude.

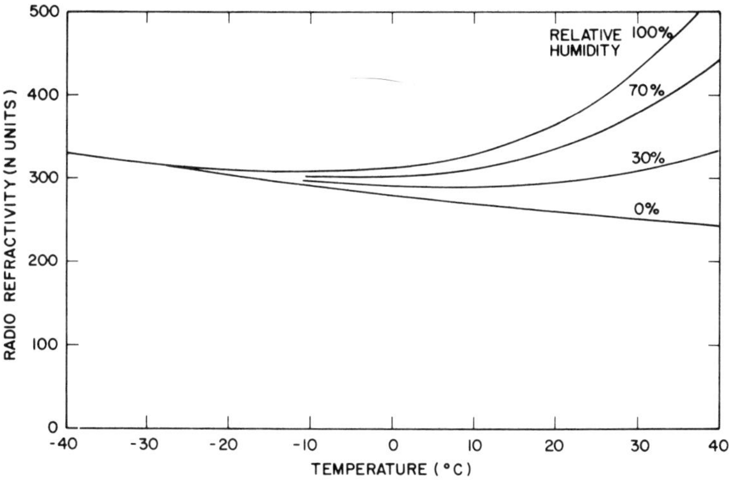

Fig. 3.2 The radio refractivity of air at a pressure of one atmosphere for four values of relative humiditiy. As the temperature rises the humidity can play an increasingly important role in determining the refractivity of the atmosphere.

water-vapor component makes a comparatively negligible contribution to the refractivity, so our experience with the atmospheric bending of light rays generally does not apply in the radio domain.

Now the deviations from rectilinear propagation will be determined by the *variations* in radio refractivity in the atmosphere through which the wave travels. The dominant variations will occur in the vertical direction with the pressure, and usually with the temperature and vapor pressure of water, decreasing with increasing height above the ground. If the atmosphere is well mixed, that is to say the air mass has been thoroughly mixed by convection, eddy turbulence, and molecular diffusion and if it is in a state of mechanical equilibrium (gravitational and buoyant forces balanced), then we can conveniently specify the variation in P, T, and with height. Let us describe the water-vapor content in a well-mixed atmosphere by the *specific humidity* α, where

$$\alpha = \frac{M_w}{M + M_w} , \qquad (3.2)$$

and M_w and M are, respectively, the mass of water vapor and the mass of dry air in a given volume. Figure 3.3 shows a family of curves that specify the refractivity as a function of elevation for several values of specific humidity within the range frequently encountered in practice.

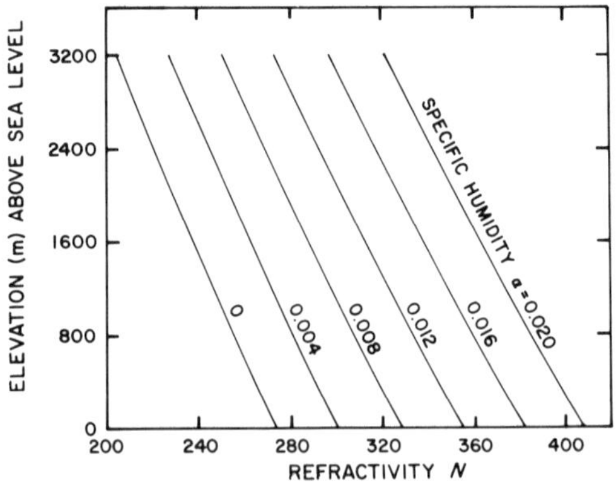

Fig. 3.3 Variation of refractivity with elevation in well mixed atmospheres for several values of specific humidity. The vertical gradient of the refractivity dN/dh is approximately $-0.023\ M^{-1}$ throughout the range of h and α represented here.

Figure 3.3 reveals three significant facts concerning the variation of N with elevation in a well-mixed atmosphere:

(a) The slope dN/dh, known as the *refractivity gradient* has nearly the same numerical value, —0.0082 N/ft or —27 N/km, throughout the range of values of h and α in the figure. This property provides the basis for a simple and convenient method for tracing the path of a wave front through a well-mixed atmosphere.

(b) The curves of refractivity *vs.* height are not quite straight lines, but dN/dh becomes slightly less negative as height increases. However, this curvature is not large enough to affect the validity of the standard ray-tracing method.

(c) The radio refractivity N is a strong function of α and increases rapidly as the specific humidity increases (see also Figure 3.2). Hence, we can expect large changes in refractive effects in cases where the atmosphere ceases to be well mixed and the temperature is high enough for the air to contain an appreciable amount of water vapor (see Figure 3.1).

In all cases there is an upper limit to the height of well-mixed regions. The upper limit occurs when the temperature reaches the saturation value for the particular value of the specific humidity. This upper bound can be recognized visually as the height at which clouds appear.

Because the refractivity N decreases with height, as Figure 3.3 shows, we expect qualitatively that the effect of refraction will be to bend horizontal rays downward and to carry the radio waves to some extent around the curved earth. Rather than to repeat the detailed analysis of the ray-tracing problem, which must be solved in this case, we shall state the results of this analysis. [The reader is referred to Livingston (B 1970), Chapter 4, for a clear exposition.]

It turns out that the refractivity gradient dN/dh causes the nearly horizontal rays to be bent with a radius of curvature ρ given by

$$\rho = \frac{n}{-\dfrac{dn}{dh} \cos \psi} \tag{3.3}$$

ρ where ψ is the angle that the ray makes with the horizontal. It was seen in Figure 3.3 that dN/dh is nearly constant in a well-mixed atmosphere, and so from Equation (3.1a) the quantity dN/dh must therefore be effectively constant. It follows then that the radius of curvature ρ must be constant for all rays close enough to the horizontal that $\cos\psi \approx 1$. Hence the rays in which we are interested will be arcs of a circle of constant radius ρ. This circumstance allows us to make use of an ingenious geometrical transformation: the actual radius of the earth R_e (6,370 km) is replaced by an effective radius KR_e such that we can *represent the refracted rays as straight lines*. The factor K which accomplishes this transformation is given by

$$K = \frac{\rho}{\rho - R_e} \quad . \tag{3.4}$$

Now in terms of the vertical gradient of index of refraction n (or radio refractivity N), we can express K as

$$K = \left(1 + \frac{R_e}{n} \frac{dn}{dh}\right)^{-1} = \left(1 + 10^{-6} R_e \frac{dN}{dh}\right)^{-1} \tag{3.5}$$

Here we have made use of the fact that n is very nearly equal to one as can be seen from Figure 3.2. For a well-mixed atmosphere, the refractivity gradient is about —27 N/km and K = 1.2. It has been customary, however, to adopt K = 4/3 as a standard working value, giving rise to the term, *4/3 - earth atmosphere*. But the well-mixed atmosphere would be a 6/5 - earth atmosphere. In practice, however, radiosonde meaurements of atmospheric structure show that K can assume a variety of numerical values. The appropriate value of K will depend on geographical location, and with changing weather conditions at a given location, one can expect appreciable day-to-day changes in K as well.

Of course, atmospheric structure is by no means limited to cases in which the refractivity gradient is constant. Temperature inversions and humidity lapses may occur to produce highly variable refractivity profiles. For example, evaporation from the surface of the sea at low latitudes may lead to comparatively large values of the specific humidity in a shallow layer near the sea surface. Referring to Figure 3.3 we can understand how this mechanism could produce large values of N near the surface with an otherwise drier well-mixed atmosphere above. If in this example the magnitude of the refractivity gradient becomes large enough, Equation 3.3 shows that the radius of curvature ρ may become sufficiently small for a ray to be bent around the curve of the earth or even down to reflect forward from the surface of the sea. This process occurs to produce a so-called *evaporation duct* in which radio waves can propagate beyond the horizon. McCue (R 1978) has given a complete account of this phenomenon and its implications for low-altitude propagation near Kwajalein Atoll.

Many other atmospheric processes can lead to ducting and greatly extended propagation. Basically, ducting can occur when there is a transition from comparatively high refractivity near the surface to distinctly lower refractivity values above. In low and middle latitudes the air temperature, particularly in summer, may be high enough for the water-vapor component to play an important role (see Figure 3.2). However, when the temperature is low, the atmosphere can contain only a small amount of humidity even when saturated. Hence in winter at higher latitudes, ducting must be a consequence of temperature inversions.

There are then at least two situations in which extended detection ranges over continental land masses may be expected because of refraction effects. After rain the evaporation of water on the ground can produce an over-land evaporation duct. This phenomenon is, in fact, observed by weather radars (personal communication, R.K. Crane). When the ground is wet and when there is little wind, ground clutter is found to spread out to much greater ranges. A totally different effect may be expected on clear nights over deserts or snow-covered terrain. Radiational cooling at the surface can produce a strong temperature inversion near the ground and, therefore, anomalous propagation (Nottarp, J 1967). This phenomena will be restricted to clear nights when there is no wind to mix the air near the ground. However, ducting may be less effective in increasing detection ranges over land than over water because diffuse scattering and absorption will be greater at the ground or snow surface than at the surface of a calm sea or lake. Both scattering and absorption by the land surface tend to remove energy from the duct.

The widely accepted procedure for characterizing the effects of atmospheric refraction on radio propagation in the design of ground-to-ground communication systems is to use balloon-borne radiosonde measurements to estimate the change ΔN in refractivity between the ground and an altitude of 100 m above the ground. This value ΔN for the first 100m of height is then used to estimate dN/dh in Equation (3.5) and, thereby, to calculate K. The deficiencies in this procedure are mainly the result of the fact that a radiosonde is not designed to make accurate low-altitude measurements. However, one can obtain some insight into the variability of refraction effects at various geographical locations because the radiosonde soundings are usually made at 12-h intervals for meteorological purposes. The distribution of the values of $\Delta N/\Delta h$ and K for sites in Canada and the northern United States has been published by Segal and Barrington (R 1977), and selected worldwide measurements of refractivity gradients have been published by Samson (R 1976) for use in designing line-of-sight communications systems.

Figure 3.4 shows two examples of the probability distributions of ground-based refractivity gradients and K values based on radiosonde measurements for Edmonton, Alberta, and Inuvik, Northwest Territories, taken from the report of Segal and Barrington (R 1977). These plots show separately the distributions for three-month intervals starting in January. Note that the values of K, reading downward in these plots, become increasingly large and change over to negative numbers. This happens because as $\Delta N/\Delta h$ becomes increasingly negative in Equation (3.5), the expression in parentheses goes to zero, making $K = \infty$ when $\Delta N/\Delta h = -157$ N/km. For this value of the refractivity gradient, the ray curvature becomes equal to that of the earth. Even more negative values of $\Delta N/\Delta h$ produce negative values of K, thus producing situations in which ducting would be expected to occur.

From the examples in Figure 3.4 we can obtain estimates of the frequency of occurrence of various values of K. At both Edmonton and Inuvik the values of K lie between 1 and 2 during 80% or more of the year. At Edmonton, K exceeds two or becomes negative (ducting condition) most frequently during July-September when these conditions occur about 20% of the time. At Inuvik the same conditions for extended propagation range ($K > 2$ or $K < 0$) occur during January-March when the frequency of occurrence is about 20%. Abnormally shortened ranges ($0 < K < 1$) are relatively uncommon in these examples, being present 5% of the time or less at both sites.

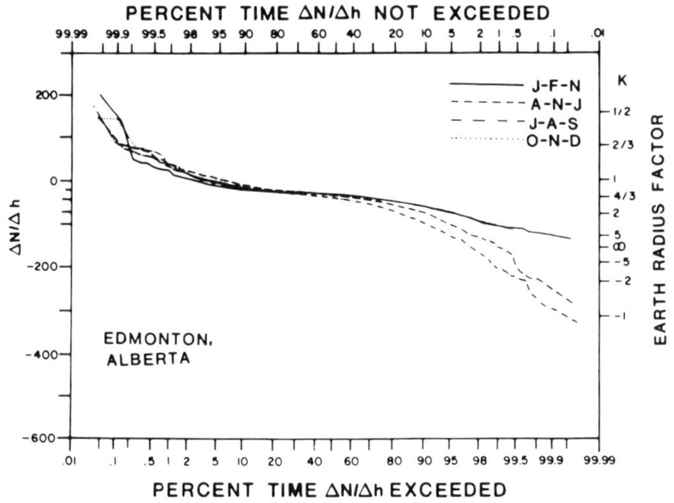

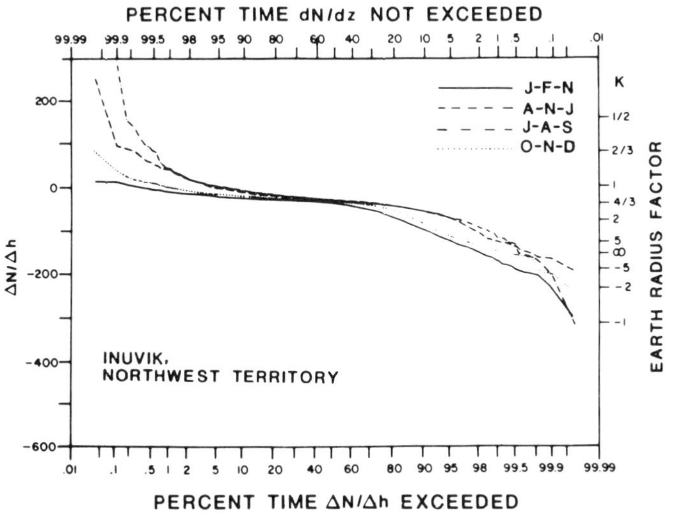

Fig. 3.4 Two examples showing the seasonal frequency of occurrence of refractivity gradients and K values. The refractivity gradients for these Canadian sites in N-units/km were determined for the lowest 100-m layer of the atmosphere from radiosonde measurements. These curves were taken from Segal and Barrington (R 1977).

Up to this point we have not considered the absorption effects of rain and atmospheric gases on the propagation of radio waves. Measurable rainfall occurs less than 5% of the time except in coastal areas (see Samson, R 1976), and rain attenuation has not been included within the scope of this report.

A convenient summary of the phenomena of rain attenuation has been given by R.K. Crane in Section 2.4 of Meeks (B 1976). Figure 3.5 shows plots of the attenuation one-way in dB/km as a function of frequency for rain (three different rain rates), for typical clouds, and for fog. For radio frequencies VHF through X-band, the droplets are small compared with wavelength λ, and the scattered power is proportional to λ^{-4} (Rayleigh scattering). At shorter wavelengths when the droplet circumference becomes comparable with or larger than λ, the scattering becomes proportional to the geometrical cross section of the droplet and is nearly independent of λ as we see in Figure 3.5.

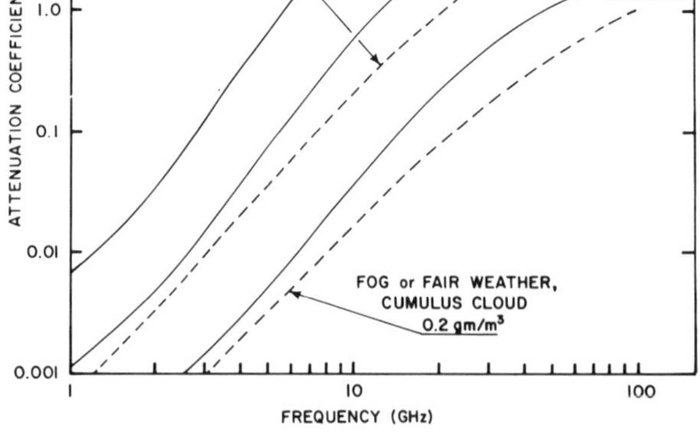

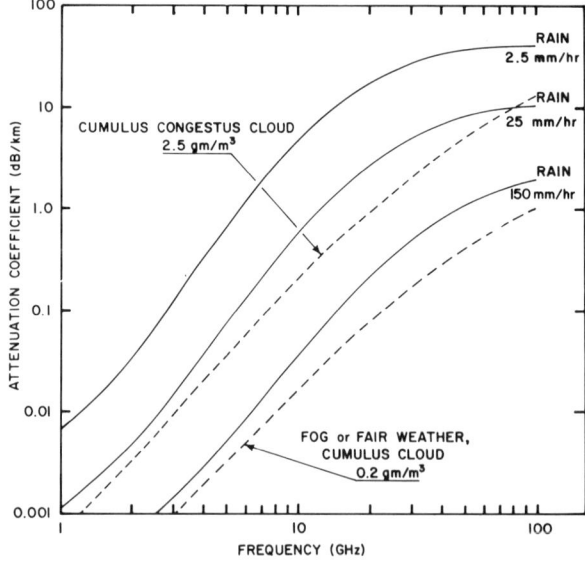

Fig. 3.5 Attenuation coefficient *vs.* frequency for rain, clouds, and fog.

Absorption by atmospheric gases is negligible at frequencies below about 10 GHz compared with the uncertainties of low-angle propagation even in the most predictable situations. However, significant amounts of absorption may occur at frequencies above 10 GHz. Absorption by atmospheric gases in the frequency range from 10 to 300 GHz is thoroughly discussed by J.W. Waters in Section 2.3 of Meeks (B 1976).

4. REFLECTION AND ABSORPTION EFFECTS OF TERRAIN

4.1 Introduction

In studying the propagation of radio waves at low-elevation angles over the surface of the earth we must take into account forward scattering (or multipath effects). We consider first reflections from idealized dielectric surfaces in the form of smooth planes (Section 4.2). Such surfaces produce only coherent reflections. Then we consider the more complicated problem of reflections from rough surfaces where both coherent and diffuse scattering can occur (Section 4.3) Finally, we discuss briefly absorption effects of trees and other vegetation which may cover terrain.

4.2 Reflection Coefficient for a Smooth Plane Surface

When a wave encounters a smooth plane surface which is very large in extent compared with a wavelength, specular reflection takes place, and the angles of incidence and reflection are equal (Snell's law). For our propagation geometry the waves move in a nearly horizontal direction, and we shall use the common terminology as follows: vertical polarization refers to linear polarization with the electric field vector lying in the vertical plane containing the incident and reflected rays, and horizontal polarization refers to linear polarization with the electric field horizontal.

The reflection coefficient is defined as the ratio of the amplitude of the reflected wave to the amplitude of the incident wave. The classical formulas derived first by Fresnel in 1816 give the reflection coefficients for horizontal and vertical polarization as a function of the electrical properties of the ground substance.

The relevant properties of matter are specified by a complex number, the relative dielectric constant ε_r, which is the ratio of the dielectric constant of the material to the dielectric constant of a vacuum. The complex constant ε_r may be written as

$$\varepsilon_r = \varepsilon_1 - j\,\varepsilon_2 , \qquad (4.1)$$

where the imaginary part ε_2 is zero when there are no losses in the material. Alternatively, the imaginary part may be expressed in terms of the conductivity σ in mhos/meter and the wavelength λ in meters as

$$\varepsilon_2 = 60\,\lambda\,\sigma . \qquad (4.2)$$

However it should be remembered that ε_1 and σ are functions of frequency, and these constants are not equal to the dielectric constant and conductivity when the electric field is constant. In general the values of these parameters are temperature dependent as well as frequency dependent, and the constants must be determined by laboratory measurements. The reflection coefficient Γ is also a complex quantity

$$\Gamma = \rho e^{-j\phi} , \qquad (4.3)$$

where ρ is the amplitude ratio and ϕ the phase shift, defined here as a phase lag of the reflected wave with respect to the incident wave.

Rather than repeat the Fresnel equations here, we shall present only the computed reflection coefficients as a function of angle in graphical form for representative types of ground and for sea and lake water. The equations may be found in Beckmann and Spizichino (B 1963), pp. 218-219, or in Kerr (B 1951), pp. 296-403. A computer subroutine incorporating the Fresnel equations is given in Appendix A as subroutine FRESNL.

The dielectric constant of soil is primarily determined by moisture content. Sand and soil with high clay content have nearly the same dielectric constant when dry. Wang (R 1979) discusses in detail the dielectric properties of soils and soil-water mixtures. In Table 4.1 we list values of ε_1 and ε_2 for soil at a representative set of frequencies and for moisture contents from 0.3 to 30% by volume. The values of ε_1 and ε_2 in this table are those given by Njoku and Kong (J 1977). Note that the dielectric constants in Table 4.1 are only significant for 20

*Values derived from laboratory measurements at a temperature of 21°C, reported by Njoku and Kong (J 1977).

and 30% moisture content where water rather than soil produces the frequency dependence. Figure 4.1 shows the reflection coefficient for four different soil-water mixtures at a frequency of 8 GHz. The amplitude ratio ρ is plotted as a function of grazing angle, which is the angle between the incident ray and its projection on the horizontal reflecting plane. Note that the value of ρ is different for horizontal and for vertical polarization. The value of the phase-shift ϕ is also plotted as a function of grazing angle in Figure 4.1 for vertical polarization. For horizontal polarization ϕ is nearly equal to π over the entire range of grazing angles.

The reflection parameters for snow, ice, and water show qualitatively similar behavior as a function of grazing angle for various values of the surface-material parameters. We now summarize the characteristics of the reflected waves using Figure 4.1 as an example. (*a*) The magnitude of the *reflection coefficient for horizontal polarization* ρ_h has the value 1 at very small grazing angles and decreases monotonically as the grazing angle ψ increases. (*b*) The *coefficient for vertical polarization* ρ_a on the other hand has a single, well-defined minimun, which is $\rho_V = 0$ provided $\sigma = 0$. The corresponding grazing angle for a loss-free dielectric is given by $\sin^{-1}(1/\sqrt{\varepsilon/\varepsilon_0})$, and the complement of this angle is called Brewster's polarizing angle.*

Table 4.1 Electromagnetic Properties of Soil*

Frequency (GHz)	Moisture Content by Volume: 0.3 ε_1	0.3 ε_2	10% ε_1	10% ε_2	20% ε_1	20% ε_2	30% ε_1	30% ε_2
0.3	2.9	0.071	6.0	0.45	10.5	0.75	16.7	1.2
3.0	2.9	0.027	6.0	0.40	10.5	1.1	16.7	2.0
8.0	2.8	0.032	5.8	0.87	10.3	2.5	15.3	4.1
14.0	2.8	0.035	5.6	1.14	9.4	3.7	12.6	6.3
24.0	2.6	0.030	4.9	1.15	7.7	4.8	9.6	8.5

*This effect can be used to advantage in the design of airborne radars which look down for low-flying targets. By employing vertical polarization such radars can eliminate or greatly reduce multipath due to ground reflections.

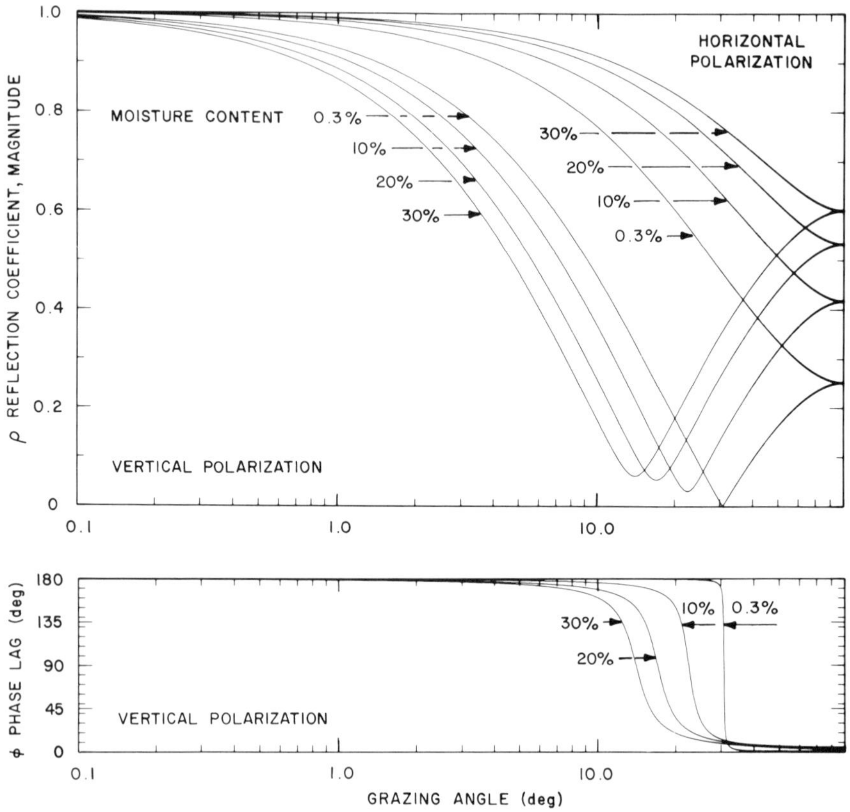

Fig. 4.1 The reflection coefficient as a function of grazing angle for four soil-water mixtures at a frequency of 8 GHz.

(*c*) At normal incidence (grazing angle 90°) the reflection coefficients for horizontal and vertical polarization are equal; the Fresnel equations reduce to

$$\Gamma = \frac{1 - \sqrt{\varepsilon_r}}{1 + \sqrt{\varepsilon_r}}$$

in this case. (*d*) For *horizontal polarization* the *phase lag* Φ is nearly π for all values of ψ, but for vertical polarization the phase lag Φ goes from π at small grazing angles ($\psi \leq 8°$) to zero for large grazing angles ($\psi \geq 45°$) with the changeover occurring around Brewster's angle. (*e*) For very low grazing angles in the neighborhood of 1° or less, both ρ_h and ρ_v are nearly one and Φ_h and Φ_v are nearly π. As a result, there should be *little difference* in the *propagation of horizontally* and *vertically polarized waves* over the ground at very low grazing angles. The conclusion is borne out by many propagation experiments, for example LaGrone (J 1960), McPetrie and Ford (J 1964), and Oxehufwud (J 1959). (*f*) The electromagnetic properties of the various ground types all lead essentially to the same reflection coefficient at very low grazing angles as Figure 4.1 shows.

The reflection properties of ice and snow have been described by Evans (J 1965) and by Linlor (J 1980). For ice the real part of the relative dielectric constant is 3.2 for frequencies between 100 MHz and 20 GHz, and the conductivity σ is 5.7×10^{-5} mhos/m for frequencies from 100 MHz to 2 GHz. Above 2 GHz the value of σ increases and is equal to 6×10^{-4} at 10 GHz. The dielectric constant of snow depends on its density, and if snow is dry, ε_1 is roughly equal to 3.2 times the ratio of the density of the snow to the density of ice. If the snow is wet, the value of ε_1, may be increased somewhat over that of dry snow because of the presence of liquid water. It is interesting to note in Table 4.1 that the dielectric constant of ice is roughly equal to that of dry soil (desert), and for snow ε_1 will be somewhat smaller. Hence reflections from flat desert surfaces will be nearly the same as from flat surfaces of ice or dense snow provided that the wave that propagates into the snow is not appreciably reflected from the ground underneath. In the special case of a uniform layer of snow on flat ground, the reflection coefficient depends on interference effects between the waves reflected from the two surfaces. A solution to this problem for two or more layers is given by Kong (B 1975), pp. 112-127. However in most cases the snow depth will not be uniform, and the solution to the simplified problem cannot be used directly.

The dielectric properties of water depend on frequency, temperature, and the presence of impurities. Over the frequency range from 100 MHz to 8.0 GHz the dielectric constant of water can be calculated as a function of salinity and temperature from a model developed by Klein and Swift (J 1977). We have used this model to calculate representative values of ε_1 and σ for lake water (salinity 0.4%) and for sea water (salinity 3.5%). Table 4.2 gives the values for lake water, and Table 4.3 gives the values for sea water. Both of these tables show little change in the dielectric properties over the frequency range from 100 MHz to 2 GHz. At higher frequencies there is

Table 4.2 Electromagnetic Properties of Lake Water*

Frequency (GHz)	T = 0°C		T = 10°C		T = 20°C	
	ε_1	σ	ε_1	σ	ε_1	σ
0.1	85.9	0.38	83.0	0.51	79.1	0.64
1.0	84.9	0.87	82.5	0.84	78.8	0.88
2.0	82.1	2.3	81.1	1.8	78.1	1.6
3.0	77.9	4.4	78.9	3.4	76.9	2.7
4.0	72.6	7.0	75.9	5.5	75.3	4.3
6.0	61.1	13.0	68.7	11.0	71.0	8.3
8.0	50.3	18.0	60.7	16.0	65.9	13.0

*These values were computed from equations given by Klein and Swift (J 1977); a salinity of 0.4% was assumed. Values of σ are in mhos/m.

Table 4.3 Electromagnetic Properties of Sea Water*

Frequency (GHz)	T = 0°C		T = 10°C		T = 20°C	
	ε_1	σ	ε_1	σ	ε_1	σ
0.1	77.8	2.9	75.6	3.8	72.5	4.8
1.0	77.0	3.3	75.2	4.1	72.3	5.0
2.0	74.6	4.6	74.0	5.0	71.6	5.6
3.0	71.0	6.4	72.1	6.4	70.5	6.7
4.0	66.5	8.8	69.5	8.2	69.1	8.0
6.0	56.5	14.0	63.2	13.0	65.4	12.0
8.0	47.0	19.0	56.2	18.0	60.8	16.0

*These values were computed from equations given by Klein and Swift (J 1977); a salinity of 3.5% was assumed. Values of σ are in mhos/m.

a significant decrease in ε_1, and an increase in σ. Figure 4.2 shows the reflection coefficient of lake water (temperature 10° C and salinity 0.4%) as a function of grazing angle for three representative frequencies: 100 MHz, 1 GHz, and 8 GHz. This figure shows significant differences in the reflection coefficient between horizontal and vertical polarization even at low grazing angles. The reflection coefficients shows little frequency dependence at grazing angles less than 3^0. At larger grazing angles the depth of the minimum in ρ_h is determined by the size of ε_2, which is 91.8, 15.1, and 36.0 for the frequencies 0.1, 1.0, and 8.0 GHz, respectively. Figure 4.3 shows a plot of the reflection coefficient of sea water (temperature 10° C and salinity 3.5%) for the same three frequencies represented in Figure 4.2. For sea water the reflection coefficient exhibits a stronger frequency dependence than for lake water as Figures 4.2 and 4.3 show.

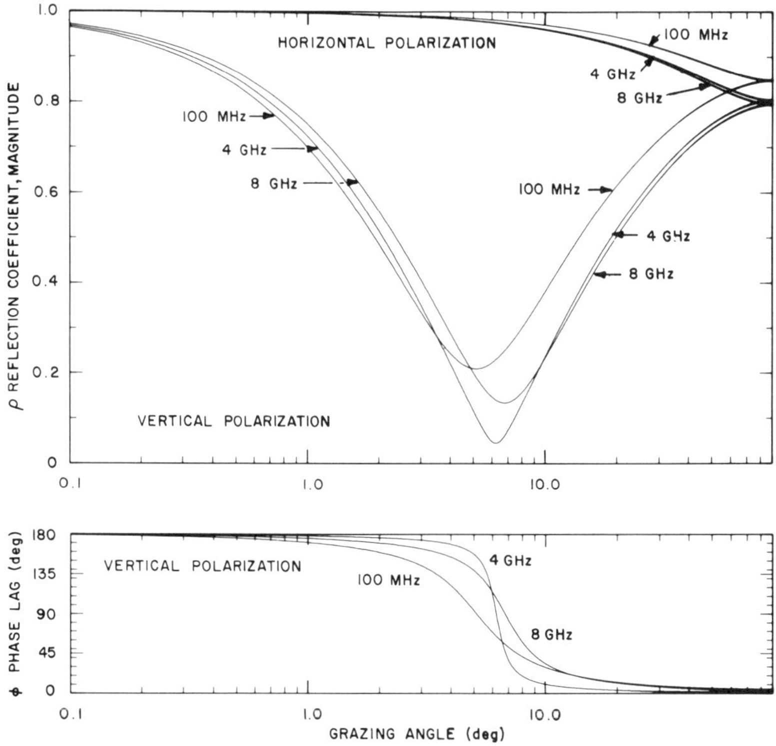

Fig. 4.2 The reflection coefficient as a function of grazing angle for lake water. Separate curves are shown for frequencies of 100 MHZ, 4 GHz, and 8 GHz; a temperature of 10° C and a salinity of 0.4% are assumed.

This is a result of the higher conductivity of sea water. The imaginary part of the complex dielectric constant, according to Equation (4.2) and Table 4.3, increases as the frequency decreases, and the minimum in ρ_V shifts to lower angles and becomes less pronounced. Also the phase lag ϕ makes a more gradual change from π as the grazing angle increases.

The special case of smooth reflecting surfaces discussed in this section can be used to represent actual terrain reflectivities in at least three cases: (*a*) ocean or lake surfaces with negligible wave disturbance, (*b*) flat desert surfaces, and (*c*) snow-covered plains. Nevertheless, in the great majority of cases actually encountered, the ground is not smooth, and we must find ways to take into account not only the gross terrain surface features but also the small-scale local roughness and vegetative cover.

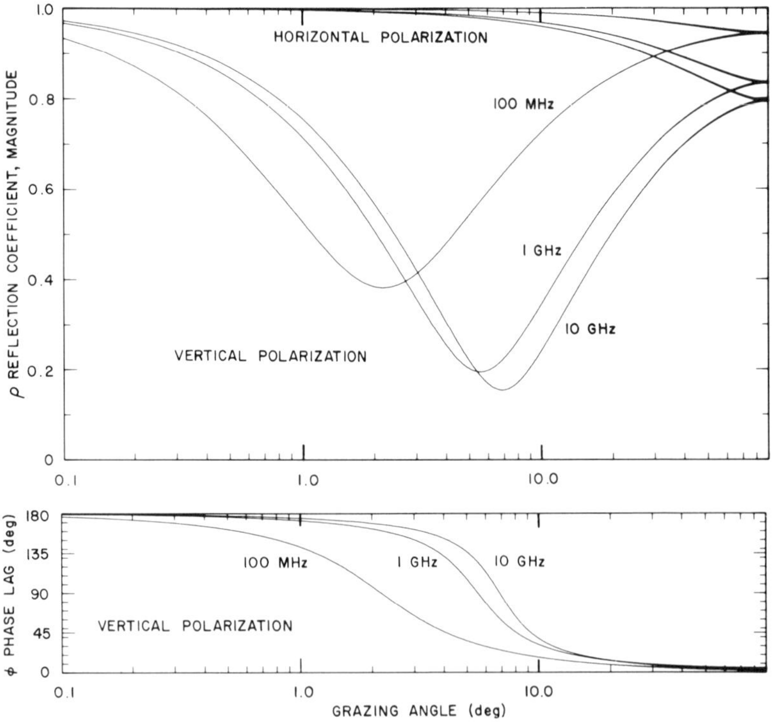

Fig. 4.3 The reflection coefficient as a function of grazing angle for sea water. Separate curves are shown for frequencies of 100 MHz, 1 GHz, and 8 GHz; a temperature of 10° C and a salinity of 3.5% are assumed.

4.3 Reflections from Rough Surfaces

When an electromagnetic wave propagates at low angles, the direct wave combines with waves reflected from the ground so that the target is illuminated by a complex wave front. We now consider the nature of ground reflections that must be taken into account if we are to predict propagation under realistic conditions, that is over a rough terrain surface. Let us think of the terrain as a large collection of scatterers distributed so as to represent the reflection properties of the surface features. If we examine the phase of the individual waves arriving at the target from these scatterers, we will generally find that a fraction of the phases will be found to be randomly distributed (*diffuse scattering*) and that the remainder will be strongly correlated in phase (*coherent scattering*). In the discussions that follow we shall neglect the diffuse scattering. We do so because the signals from the diffuse scatterers combine with random phases and add up to produce a field strength that is small compared with the direct wave. Although in many cases diffuse scattering will affect the angular accuracy of tracking radars (Mrstik and Smith, J 1978, and White, J 1974), the effects on the pattern-propagation factor and the detection performance of a radar are negligible. On the other hand, coherent scattering, depending on the nature of the terrain, can produce a reflected wave comparable in amplitude with the direct wave. (In fact, if the terrain should happen by chance to have a surface of nearly spheroidal contour, the reflected waves could be brought to a focus at the target, producing a reflected wave much stronger than the direct wave.) Coherent reflection may or may not be produced depending on the roughness of the terrain and the wavelength of the signal, but if coherent reflection occurs, the reflected wave will propagate in the same azimuthal direction as the direct wave. On the other hand, diffuse reflections will scatter energy in all directions. In the backward direction the diffuse reflection of course produces the backscattered ground clutter seen by the radars.

The characteristics of specular and diffuse reflection are well described in the following paragraphs from Beckmann and Spizzichino (B 1963), page 241:

> Specular reflection is a reflection of the same type as caused by a smooth surface: it is directional and obeys the laws of classical optics. Its phase is coherent and it is the result of the radiation of

the points on the Fresnel ellipse, which transmit waves of approximately equal phases toward the receiver. Its fluctuations have a relatively small amplitude.

Diffuse scattering is a phenomenon that has little directivity and which consequently takes place over a much larger area of the surface than the first Fresnel zone. Its phase is coherent and its fluctuations, which have a large amplitude, are Rayleigh-distributed.

The existence of these two components of the reflected field is well established by optical as well as radio experiments.

A simple theoretical model has been developed to calculate the *specular reflection coefficient*, R_s, of rough terrain. Because of the assumed randomness of the terrain, R_s must itself be random with a RMS value given by

$$(R_s)_{RMS} = (\rho_s)_{RMS}\, \Gamma \,. \tag{4.3a}$$

Here ρ_s is the *specular scattering coefficient* and Γ is the *reflection coefficient* for a smooth-plane earth. For a Gaussian-model rough surface, according to Beckmann and Spizzichino (B 1963), page 246, we have

$$\langle |\rho_s|^2 \rangle_{\text{average}} = e^{-(\Delta\Phi)^2} \tag{4.4}$$

with

$$\Delta\Phi = \frac{4\pi\, \Delta h \sin\psi}{\lambda} \,. \tag{4.5}$$

The model surface is specified by the single parameter, Δh, which is the standard deviation of the normal distribution of heights measured from a plane surface, ψ is the grazing angle, and λ is the wavelength. The quantity $\Delta\Phi$ is just the phase difference between two rays that reflect from the rough surface: one ray reflecting from the mean surface level and the other from a bump of height Δh on the surface.

This simple theoretical model agrees fairly well with actual measurements (see Beckmann and Spizzichino, B 1963, pp. 303 and 317) over a range of grazing angles that in some cases goes below 1°. However, the propagation measurements with which the model can be compared permit only relatively crude estimates of Δh, and it does not appear useful to consider more elaborate models until more complete and accurate propagation data with quantitative ground truth are obtained.

Notwithstanding the limitations of the Gaussian terrain model, it is useful to examine those combinations of radio wavelength and terrain type for which this model predicts appreciable coherent reflection. Let us consider an idealized case of propagation over a spherical earth with homogeneous surface roughness specified by Δh. Figure 4.4 shows a diagram in which Δh and λ appear as the x- and y-axes. Also indicated in this figure are the terrain-type and sea-state designations that roughly correspond to various ranges of values of Δh. These terrain-roughness designations are the ones used by Longley and Rice (R 1968). We have fixed values of ρ_s and ψ and have plotted on the diagram the locus of $(\Delta h, \lambda)$ values that would give rise to specific values of scattering coefficient at specific grazing angles. These loci from Equations (4.4) and (4.5) must appear on the diagram as diagonal straight lines, all with the same slope. We represent two cases in Figure 4.4. For both cases we assume nominal values for the height of the radar antenna (30 m) and for the altitude of the target above ground (90 m). At a comparatively short range of 10 km over a spherical earth, the grazing angle of the reflected ray is 0.7°, and at a longer range of 30 km the grazing angles $\Gamma \approx 1$ (see Section 4.2). These bands represent transition plotted two bands with the left edge of each band corresponding with $\rho_s = 0.9$ (largely specular reflection) and the right edge with $\rho_s = 0.3$ (largely diffuse reflection). The reflection coefficient R_s in Equation (4.3) is, of course, determined by ρ_s because at these low grazing angles $\Gamma = 1$ (see Section 4.2). These bands represent transition regions betwen the upper-left region in Figure 4.4, where the terrain reflection is specular and the lower-right region where the reflection is entirely diffuse. For the shorter range of 10 km, this model predicts that high reflectivity would be expected only for low VHF frequencies over smooth plains and for VHF and UHF frequencies over water. For the longer range of 30 km, the highly reflective regions are extended to include slightly rolling plains for VHF frequencies and smooth plains for UHF frequencies. The increased reflectivity for the longer range is a consequence of the factor (sin ψ) in Equation (4.5.) For all landforms with which we are concerned, excluding very unusual cases, the value of Δh will always be greater than 3 m. So, on the basis of this model, we can rule out appreciable specular reflection from terrain when the microwave bands are used (i.e., frequencies greater than 1 GHz). Nevertheless, propagation measurements even at millimeter wavelengths show the presence of significant coherent terrain reflection at grazing angles below about 1;° for example, Hayes, Lammers *et al.* (J 1979) and Haakinson, Violette, and Hufford (R 1980). Therefore we must conclude that the Gaussian terrain model breaks down for low grazing angles and high microwave frequencies because it predicts reflection coefficients that are much too small. The grazing angles and frequencies

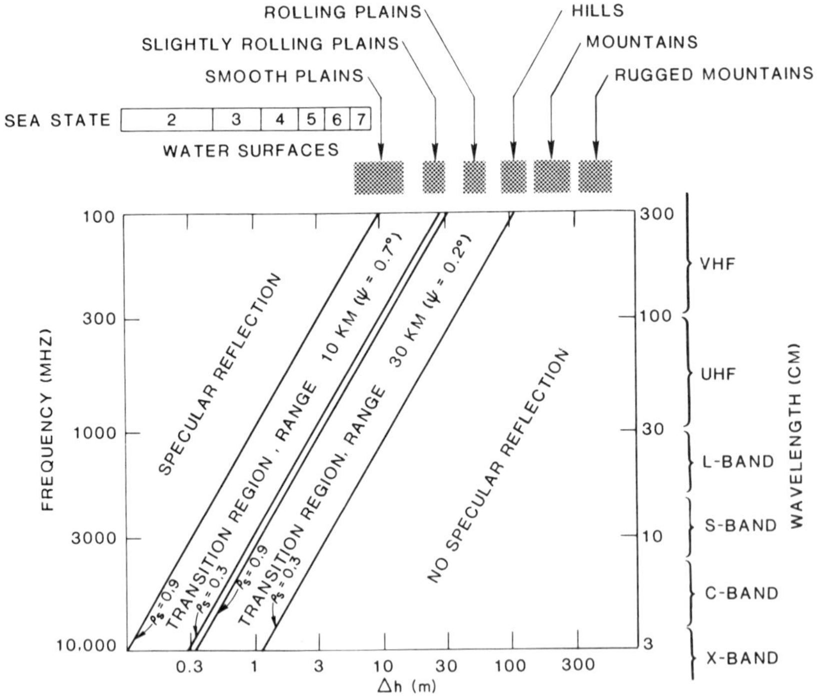

Fig. 4.4 Terrain-wavelength diagram showing the combinations of Δh and radio frequency for which appreciable specular reflection is predicted by the Gaussian terrain model. Two diagonal bands are shown representing the transition regions between specular and diffuse terrain reflection for assumed target ranges of 10 and 30 km. In both cases the antenna and target heights are 30 and 90 m.

at which large errors appear in the Gaussian terrain model are unknown. For grazing angles smaller than 1° there is insufficient experimental data to establish the reflection coeffcient as a function of frequency for any type of terrain.

In summary we can say that coherent reflection effects will be most important at VHF and UHF frequencies where the reflection coefficient may approach one in flat terrain or over water. The Gaussian model of terrain reflection gives some guidance in estimating the reflection coefficient, but the evidence shows that this model breaks down at very low grazing angles and high microwave frequencies. An extensive series of field measurements will be necessary before an adequate model can be developed.

Up to this point we have not considered specifically the effects of vegetative cover on the reflection properties of terrain. In general these effects are not well understood, but there are two groups of measurements that shed some light on the problem. At a frequency of 4 GHz we have the results of measurements made by the American Telephone and Telegraph Co. during the engineering of a transcontinental microwave relay system (Bullington, J 1954). These measurements were made over a series of paths (each roughly 25 m long) that form links in a communication chain connecting New York and Denver. In most cases, the reflection coefficients fell in the range from 0.2 to 0.4. The following paragraph from Bullington's paper describes the comparison of ground-reflection data for vegetated and barren paths.

> The data on the New York-Omaha section of the route was separated from the Omaha-Denver section to show whether or not the difference in the type of terrain adds significantly to the spread in the overall results. At first, some of the low-reflection coefficients were attributed to absorption in the numerous trees along some of the paths, but an approximately equal percentage of low reflections were found in the relatively flat, treeless areas in eastern Colorado.

At longer wavelengths, a different situation emerges. Extensive measurements have been made of television service fields at VHF and UHF frequencies. The standard receiving antenna height for these measurements is 30 ft (9 m), which is a small height compared with radar antenna heights or target heights. Nevertheless, these experimental results must be considered relevant. The following paragraphs are quoted from a survey paper by LaGrone (J 1960).

> Vegetation is an important factor in field-measured data at some television frequencies . . . At low VHF (frequencies), it was found to be negligible (in its effect), but at high VHF (frequencies), trees and tall grasses were found to absorb a significant amount of the signal.
>
> The absorption of radio waves by trees and grasses noted at higher VHF (frequencies) was found to be considerably more at UHF. Previously it was noted that trees thick enough to block vision were essentially opaque at 1000 MHz.

The comparatively small effect of vegetation on VHF propagation is qualitatively confirmed by the fact that at 67 MHz, no appreciable change in television coverage is found between summer and winter conditions in the Boston area (personal communication, Joseph Blake, Chief Engineer, WBZ-TV, Channel 4 Boston 111A.) Clearly more experimental work will be required to understand the effects of vegetative cover, particularly at UHF frequencies, for antenna (or target) heights from 100 to 500 ft (30 to 150 m), rather than 30 ft.

5. DIFFRACTION EFFECTS

5.1 Introduction

It is the nature of waves to bend around obstacles. This diffraction phenomenon plays a central role in the propagation of radio waves at low elevation angles over the ground. Hills and ridges diffract energy into the shadow zone and make possible radio communication and radar target detection within these zones. Theoretical models that describe diffraction by a knife-edge and by cylinders with various radii of curvature may be used in predicting radio propagation, and the areas of application and limitations of these simple geometrical models are fairly well understood. In this chapter we discuss these models.

In the theoretical analysis of diffraction, the obstacles are frequently chosen to be perfect conductors, and the polarization of the electromagnetic waves may be an important consideration. However, for knife-edge diffraction the results are rigorously independent of polarization for small diffraction angles, and for cylinder diffraction the polarization dependence, although strong for perfect conductors, is small for the values of dielectric constant and conductivity encountered on terrain at our frequencies of interest. In this chapter we shall assume that the polarization is horizontal (i.e., the electric vector is perpendicular to the plane containing the incident and diffracted rays). For diffraction by cylinders, Hacking (J 1970) has confirmed experimentally that vertical polarization propagates somewhat more strongly into the shadow zone than horizontal polarization, but the difference is small ($\leq$ 2 dB).

We shall consider in the following sections the Fresnel-Kirchhoff calculation of diffraction by a knife-edge and the corrections that must be added when it becomes necessary to account for the finite radius of curvature of the diffracting mask. It turns out that most hills and ridges can be well represented by a knife-edge. In the final section we discuss the problem of successive diffraction over several knife-edges.

The diffraction models which we discuss in this chapter are essentially two-dimensional models, so a question arises concerning the widths that obstruction must have perpendicular to the plane of incidence for these models to apply. The answer to this question is that the obstruction must have a transverse width greater than the first-Fresnel-zone width,

$$[\lambda d/4]^{1/2}\left[1-\left(\frac{2d_1}{d}-1\right)^2\right]^{1/2} \tag{5.0}$$

where d is the distance from the radar to the target (or transmitter to receiver), and d_1 is the shorter of the two distances: radar to obstruction and obstruction to target. For an obstruction centered on a 10-km path, the first-Fresnel-zone widths are 87, 27, and 9 m for radio frequencies 100 MHz, 1 GHz, and 10 GHz, respectively.

5.2 Diffraction by a Knife-Edge

The problem of diffraction of an electromagnetic wave by a knife-edge is a classical problem in optics solved by Sommerfield in 1896, and the detailed solution is given in Born and Wolf (B 1959), Sections 8.7 and 11.5. The geometry of the problem is illustrated in

Figure 5.1. The distances from transmitter to knife-edge and receiver to knife-edge are d_T and d_R, respectively.

Here we consider first one-way propagation from a transmitter at T to a receiver at R over a knife-edge at O. An important parameter is the clearance of the line TR over the knife-edge or, if the line TR intersects the mask, the distance from O to TR. We call this clearance distance Δ, taking Δ as positive when TR clears the mask and as negative when TR intersects the mask (R in the shadow region). Figure 5.2 shows this definition of Δ.

The solution to the knife-edge diffraction problem gives the field at R caused by a source at T, and from the principle of reciprocity the same field would be produced at T if the transmitter were placed at R. Accordingly, we consider the transmitter and knife-edge as fixed and denote the field at the receiver as $E(d_R, y_R)$. The explicit solution $E(d_R, y_R)$ for wavelength λ near the geometric shadow (for either polarization) is given by

$$E(d_R, y_R) = \frac{e^{-j(\pi/4)}}{\sqrt{2}\,r} e^{jkr}\left\{[1/2 + C(w)] + j\,[1/2 + S(w)]\right\} \tag{5.1}$$

where

$$k = 2\pi/\lambda$$

$$r = \sqrt{(y_T - y_R)^2 + (d_T + d_R)^2}$$

$$w = \sqrt{2}\left(\frac{\Delta}{\sqrt{\lambda d_T\, d_R / (d_T + d_R)}}\right) \tag{5.2}$$

$$C(w) = \int_0^w \cos\frac{\pi}{2}\tau^2\, d\tau$$

$$S(w) = \int_0^w \sin\frac{\pi}{2}\tau^2\, d\tau \quad .$$

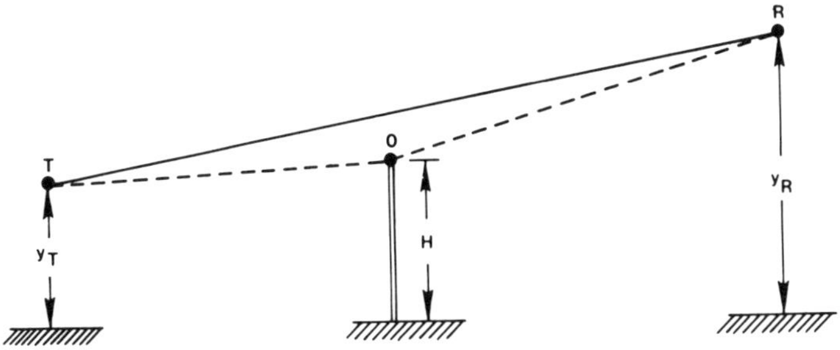

Fig. 5.1 The geometry of the diffraction problem.

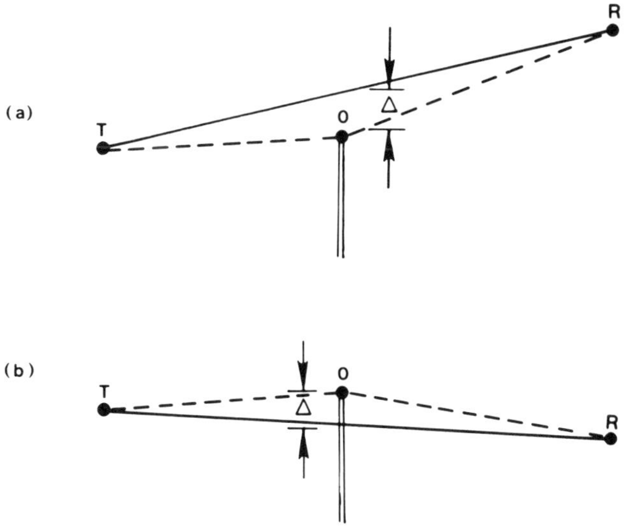

Fig. 5.2 The clearance parameter Δ. (a) An example of positive clearance, (b) negative clearance.

The functions C(w) and S(w) are the Fresnel integrals, and their argument w has a simple geometrical interpretation. When the clearance Δ has the value Δ_0 such that

$$\Delta_0 = \sqrt{\lambda\, d_T\, d_R/(d_T + d_R)} \tag{5.3}$$

then the path length TOR is longer than the direct path TR by an amount $\lambda/2$. Thus, Δ_0 as defined above is just the clearance that puts the knife-edge at the boundary of the first Fresnel zone, and from Equation (5.2) the Fresnel parameter w may be expressed as

$$\frac{w}{\sqrt{2}} = \frac{\text{clearance of ray TR}}{\text{clearance of first Fresnel zone}} = \frac{\Delta}{\Delta_0}\,. \tag{5.4}$$

In other words, $w/\sqrt{2}$ is just the clearance of ray TR expressed in units of the first-Fresnel-zone clearance. (A more detailed discussion of this problem is given in Brodhage and Hormuth, B 1977, Section 3.1.)

Now let us look at the solution to the knife-edge diffraction problem given by Equation (5.1.) Figure 5.3 shows the power propagated from T to R relative to free-space propagation. This quantity expressed in decibels is equal to 20 log F, where F is the pattern-propagation factor defined in Chapter 2.

The independent variable in this figure is just the ray clearance in units of Δ_0. Expressed in this way the result is, of course, independent of wavelength.

In order to see how radio waves of different frequencies diffract over a knife-edge, let us consider an example. Figure 5.4 shows the geometrical arrangement and a plot of the pattern-propagation factor F computed as a function of the target height for five different radar frequencies (wavelengths): 100 MHz (3 m), 300 MHz (1 m), 1 GHz (30 cm), 3 GHz (10 cm), and 10 GHz (3 cm). It is convenient to interpret the quantity F as the factor that multiplies the maximum range of the radar in free space to obtain the maximum range under propagation conditions characterized by F (see Section 2, Equation (2.3) and (2.4). Hence, a radar with a maximum range in free space of 150

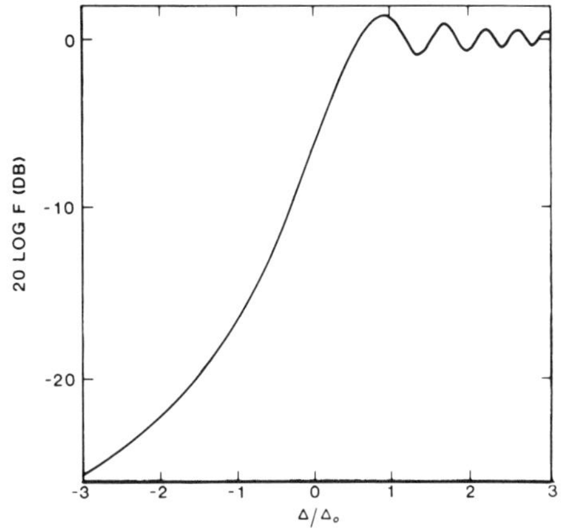

Fig. 5.3 Propagation over a knife-edge as a function of normalized clearance. The abscissa represents the clearance Δ normalized in units of first-Fresnel-zone clearance. The ordinate, 20 log F (F = pattern-propagation factor, see Chapter 2), is also the one-way propagated power relative to free space expressed in dB.

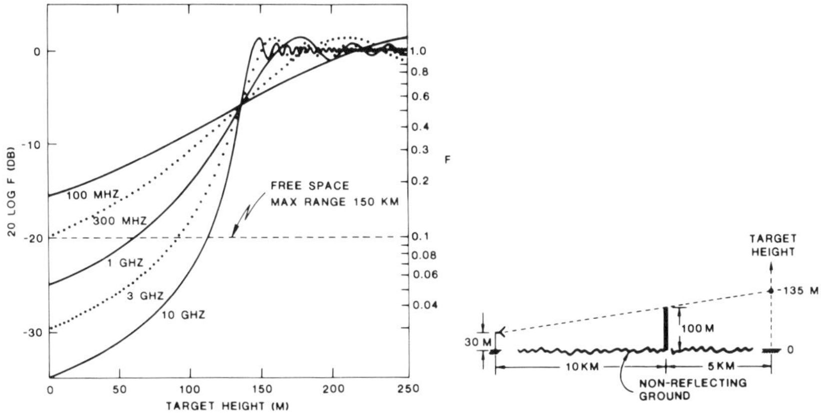

Fig. 5.4 An example of knife-edge diffraction for five radar frequencies. The abscissa of the graph represents target height as shown in the diagram. The ordinate represents both 20 log F and F, where F is the pattern-propatation factor. The dashed horizontal line shows the detection limit for targets in the shadow region. Above this line a target could be detected (in the absence of clutter) by a radar with a maximum free-space range of 150 km.

km would be able to detect the target in Figure 5.4 at a range of 15 km provided $F \geq 0.1$ in the absence of ground clutter. The curves plotted in Figure 5.4 show that the target would be detectable at any height above the ground at radar frequencies below 300 MHz. However, for higher radar frequencies — 1, 3, and 10 GHz — the minimum heights for detection are 62, 91, and 100 m above the ground, respectively. This is, of course, a simplified example that neglects ground reflection and ground clutter. In Chapter 6 we consider the combined propagation effects of diffraction and reflection.

In the design of microwave links, careful consideration must be given to the diffraction effects of terrain. Field experiments (for example, Oxehufwud, J 1959) have shown that terrain obstruction will not be evident if 60% or more of the first Fresnel zone is unobstructed. Figure 5.3 shows that for $\Delta/\Delta_o \geq 0.6$ the effect of knife-edge diffraction on one-way propagation would be no more than 1.5 dB. According to Brodhage and Hormuth (B 1976), Section 4, the design practice is to place the antennas high enough that the first Fresnel zone is entirely unobstructed in order to allow for unfavorable atmospheric refraction conditions. Depending on the refractivity gradient in the atmosphere, the earth-radius factor K (see Section 3) will determine the effective clearance. In constructing computational models of low-angle propagation, one therefore has the support of theory and practice in neglecting diffraction for cases in which $\Delta/\Delta_o \geq 0.6$. In such cases we have what is referred to as Fresnel clearance.

5.3 Diffraction by Cylinders

In some situations a more appropriate model for diffraction by a hill or ridge is provided by the case of diffraction by a cylinder. Solutions to the problem of diffraction of electromagnetic waves around a cylinder have been given by Rice (J 1954), Neugebauer and Bachynski (J 1958), and by Wait and Conda (J 1959). We shall make use of the results of Dougherty and Maloney (J 1964), who extended the solution of Wait and Conda (J 1959) and presented the results in convenient graphical form. Cylinder diffraction is presented in the form of corrections to be applied to the solution for a knife-edge. Dougherty and Maloney (J 1964) use the dimensionless parameter ρ

to characterize the effect of the finite radius of curvature of the cylinder. They define ρ as

$$\rho = \left(\frac{\lambda r^2}{\pi}\right)^{1/6} \sqrt{\frac{d_T + d_R}{d_T d_R}} \tag{5.5}$$

where r is the *radius of curvature* of the cylinder, and λ is the wavelength. Figure 5.5 shows the diffraction loss as a function of the normalized clearance Δ/Δ_o for six different values of ρ. The curve for $\rho = 0$ corresponds with the previous case of knife-edge diffraction. For a frequency of 300 MHz, for example, and a hill with radius of curvature 0.5 km forming an obstruction midway along a 10-km path, we calculate from Equation 5.5 the value $\rho = 0.131$. Referring to Figure 5.4, we can see that for $\Delta/\Delta_o = -0.5$ the one-way diffraction loss is 12 dB for a knife-edge and 14 dB for the rounded hill. So, in this case, the rounded hill gives 2 dB more loss, a comparatively small change. For shorter ranges, Equation 5.5 would give larger value of ρ and greater excess loss for an obstruction of this shape. If we go to higher frequencies (shorter wavelength) we would find little decrease in ρ because λ enters only to the power 1/6; also the dependence on r goes as the power 1/3, which means that ρ is also not a strong function of r.

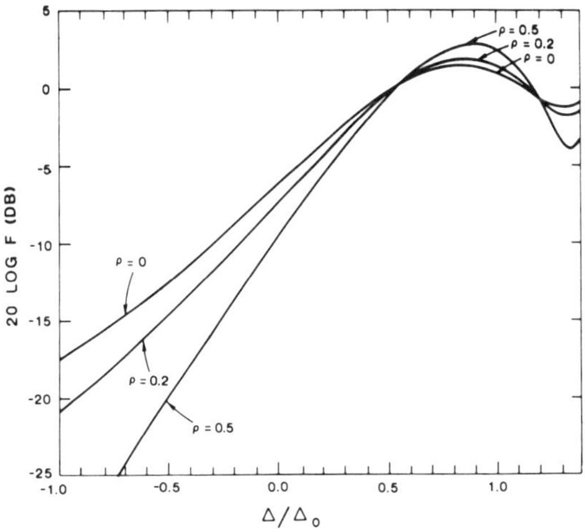

Fig. 5.5 Propagation over a cylinder as a function of normalized clearance. The parameter ρ, defined in Equation (5.5), increases with increasing radius of curvature of the cylinder. Knife-edge diffraction corresponds to $\rho = 0$.

However, when the radius of curvature begins to approach an earth radius with the corresponding $\rho > 1$, the excess loss over that of a knife-edge may become very large. This circumstance leads to the concept of *obstacle gain* whereby propagation may actually be improved by erecting a knife-edge barrier near the center of a spherical-earth path. For example, Dickson *et al.* (J 1953), it is commonly observed that very long paths over mountainous terrain provide good communication with negligible fading.

5.4 Multiple Diffraction

Radio waves propagating near the surface of the earth may frequently encounter more than one diffracting mask. The problem of multiple diffraction has been investigated by Millington, Hewitt, and Immirzi (J 1962) in the case of double knife-edges; their analysis, based on Fresnel diffraction theory, cannot easily be generalized to more than two successive knife-edges. However, Deygout (J 1966) has developed a semi-empirical procedure for predicting multiple diffraction losses which agrees with the results of Millington *et al.* for the double knife-edge and agrees well with propagation measurements over paths with up to five diffracting hills.

The Deygout method of calculation as reported in Deygout (J 1966) agrees significantly better with experiment than the methods of Bullington (J 1947) and Epstein and Peterson (J 1953).

We shall describe the Deygout method by showing an example in which three successive diffractions are handled. The method of computation can be understood as a straightforward generalization of this example. Figure 5.6(a) shows the clearances of the three masks from the line TR. These clearances are by definition negative numbers. The *principal mask* must be determined by dividing the clearance h for each mask by its first-Fresnel-zone clearance for path TR and by selecting the most negative Fresnel clearance as the principal mask. In this example M_2 is the principal mask. Next, we draw the paths to the top of the principal mask, TM_2, and M_2R, and record $\Delta_2 = h_2$ as the clearance for the principal mask. Figure 5.6(b) shows this step in the construction. Finally, we draw paths TM_1 and M_1M_2 and paths M_2M_3 and M_3R as shown in Figure 5.6(c), and the appropriate clearances Δ_1 for path TM_2, Δ_2 for path TR, and Δ_3 for path M_2R. The corresponding losses in decibels can be determined from Figure 5.3 and added together to obtain the total excess loss.

Errors in the Deygout method of calculation may be estimated by comparing the attenuations calculated for two successive knife-edges by the Deygout method and the more rigorous method of Millington *et al.* (J 1962). This comparison shows that the Deygout method overestimates the attenuation when the two knife-edges have nearly the same Fresnel-zone clearance considered separately. More specifically, errors will occur when the secondary mask has a Fresnel-zone clearance that is 70% or more of the Fresnel-zone clearance of the principal mask. If the two nearly equal masks are close together, the overestimate produced by the Deygout method may approach 6 dB. Therefore, one must be careful when employing the Deygout method in cases where two or more prominent features in the terrain are close together and have nearly the same heights.

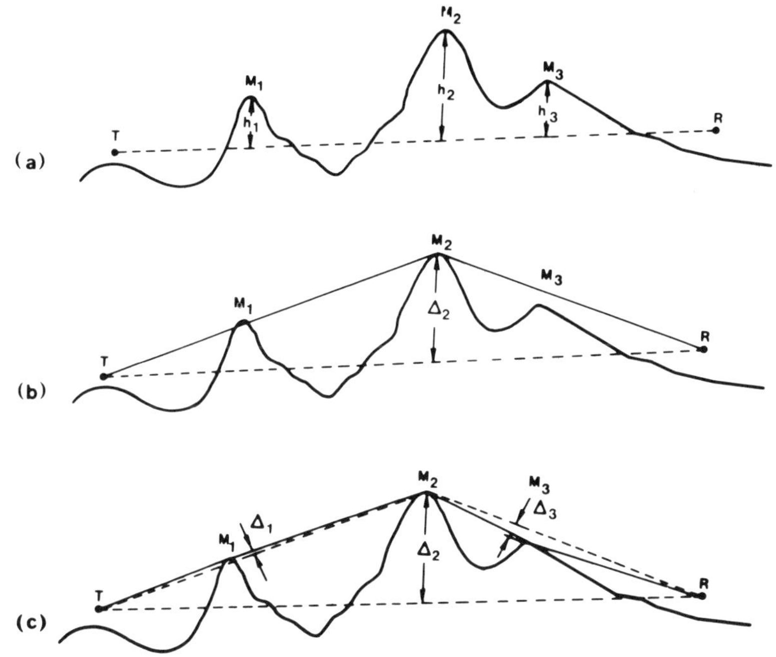

Fig. 5.6 An example of the Deygout construction. (a) The clearances from which the principal mask is determined, (b) the paths with respect to the smaller masks and (c) the clearances used to calculate the attenuations of the three masks.

6. PROPAGATION MODELS

6.1 Introduction

Here we consider the problem of computing the signal strength propagated over a specific terrain profile taking into account the effects of refraction, reflection, and diffraction. By following the discussion of refraction in Chapter 3, we see that in most cases the downward bending of ray paths by the atmosphere can be taken into account by substituting for the actual curvature of the earth a modified radius of curvature that is somewhat larger. The factor 4/3 is commonly considered as the standard correction factor, but over land this factor varies with weather conditions lying between one and two roughly 90% of the time (see Section 3). Because maps describe the terrain relief with reference to sea level, one can correct the sea level elevations for earth curvature and refraction by adding a term as follows:

$$h_i = h_i(\text{map}) - \frac{x_i^2}{2\,K\,R_e} \qquad (6.1)$$

where:

h_i	=	the height of an element of the terrain profile,
h_i (map)	=	the map height above sea level,
x_i	=	the distance from the radar to the element,
$K\,R_e$	=	the modified earth radius.

In this way the terrain profile may be obtained with the refraction correction built in.

The influence of terrain reflection is not so easily taken into account. The discussion in Chapter 4 points out that appreciable coherent reflection from the ground can be expected for terrain classified as *plains,* and primarily for radars operating in VHF or UHF frequency bands. However, the magnitude of the reflection coefficient will depend on the vegetative ground cover, and we lack an experimental basis for assigning reflection coefficients to common types of ground cover. Nevertheless, we can examine the effects of ground reflection in some particular cases by trying several values for the reflection coefficient. In all terrain types except *plains,* diffraction by hills and ridges will be the most important phenomenon governing radio propagation at low altitudes. When there is a single diffracting ridge or hill along the path, the problem is easily handled with the methods described in Section 5. If there are several diffracting masks in series along the profile, the propagation may be calculated by the Deygout method which gives a good approximation in most cases (see Section 5.4). On the other hand, over smooth plains, radio waves reflect from and diffract over what is essentially a sphere (or cylinder) with a very large radius of curvature, roughly that of the earth. This problem has been solved in general only for a smooth dielectric or conducting sphere (Section 6.4).

In the sections that follow we examine specialized propagation models: (*a*) propagation over a plane, (*b*) propagation over a knife-edge on a plane, and (*c*) propagation over a smooth spherical earth. The solutions to these problems give some insight into the problems of propagation modeling.

6.2 Propagation Over a Plane

The standard method of investigating propagation over a flat plane makes use of the fact that the plane reflects like a mirror, and the reflected wave can be assumed to originate at the mirror image of the source. In effect, one replaces the reflecting plane by the image of the source and assigns a phase shift to the image wave to take into account the phase change upon reflection. Also, the amplitude of the reflected wave must be multiplied by a factor equal to the magnitude of the reflection coefficient. The direct wave and the wave from the image combine to form a pattern of lobes and nulls. Figure 6.1 shows the geometry of the problem. If the horizontal distance to the target is large compared with the heights z_1 and z_2, then the path difference is:

$$\Delta R = d_2 - d_1 = 2 z_1 z_2 / x . \tag{6.2}$$

Let us assume that the source is a radar with the axis of its antenna beam pointed horizontally toward the target and that the pattern of the antenna is given by $f(\theta)$, where θ is the angle between the horizontal and the target viewed from the radar antenna or from the mirror image of the antenna. Then the pattern-propagation factor will be given by:

$$F = |f(\theta_1) + \rho f(\theta_2)\ e^{-j(2\pi\ \Delta R/\lambda\ +\ \Phi))}| \qquad (6.3)$$

where:

R = range from radar to target,

θ_1 = $(z_2 - z_1)/R$ (radians),

θ_2 = $(z_2 + z_1)/R$ (radians),

$\rho e^{-i\Phi}$ = complex reflection coefficient,

λ = wavelength.

Because the phase shift will normally be π at low elevation angles (see Section 4.2), nulls will be produced when the target height has values:

$$z_2 = 0,\ (\frac{\lambda x}{2z_1}),\ 2\,(\frac{\lambda x}{2z_1}),\ 3\,(\frac{\lambda x}{2z_1}) \cdots$$

and peaks produced for target heights

$$z_2 = \frac{\lambda x}{4z_1},\ 2\,(\frac{\lambda x}{4z_1}),\ 3\,(\frac{\lambda x}{4z_1}) \cdots$$

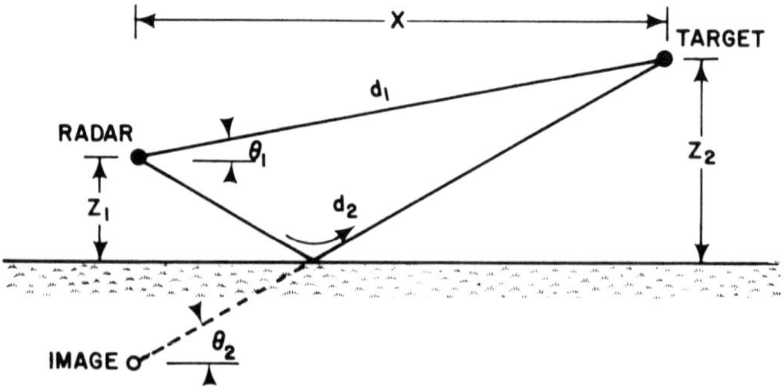

Fig. 6.1 Diagram of propagation over a plane.

The widths of the lobes are thus directly proportional to wavelength and inversely proportional to radar-antenna height.

If we neglect the antenna pattern, then the ratio of the null signal strength to the peak signal strength, expressed in decibels, is $20 \log (1-\rho)/(1+\rho)$. For example, the null-to-peak power ratios are —15 dB for $\rho = 0.7$, -9.5 dB for $\rho = 0.5$, and -5.4 dB for $\rho = 0.3$. A standard method for measuring reflection coefficient makes use of the relationship between null-to-peak ratios and reflection coefficient.

6.3 Propagation Over a Knife-Edge on a Plane

The problem of propagation over a flat plane can be generalized to include a knife-edge mask on the plane. This is accomplished by using images to represent the reflected waves. Because reflections may occur on both sides of the knife-edge, images of both the radar and the target must be considered. Figure 6.2 shows the ray paths; Figure 6.2(a) represents paths with reflections, and Figure 6.2(b) represents equivalent paths by the method of images. Four separate paths must be taken into account, and this is sometimes referred to as the four-ray problem. A program has been written to solve this

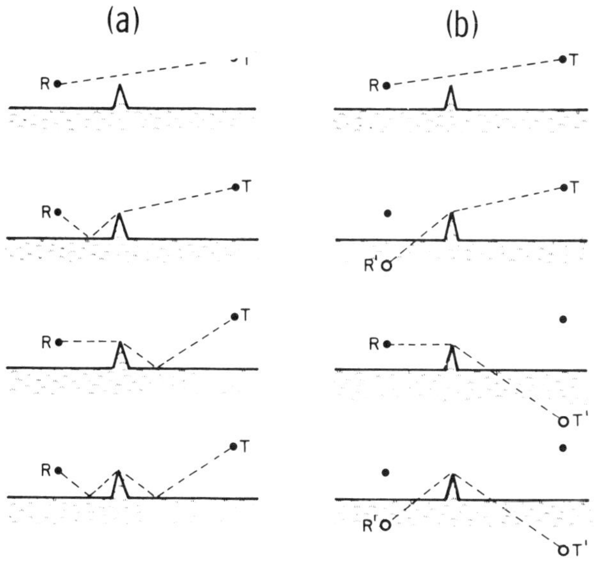

Fig. 6.2 Diagram of the four-ray problem. (a) The direct and reflected rays and (b) the equivalent construction with images.

problem when the dimensions are specified and the dielectric constants are given for the plane on each side of the mask. A complete description of this program, with a program listing, is given in Appendix A.

We have calculated the pattern propagation factor with this program for a representative situation: the target range 15 km, the mask 10 km from the radar, and both the radar and the edge of the mask are 100 m above the reflecting plane. For each radar frequency, four cases are worked out: the one-way propagation loss

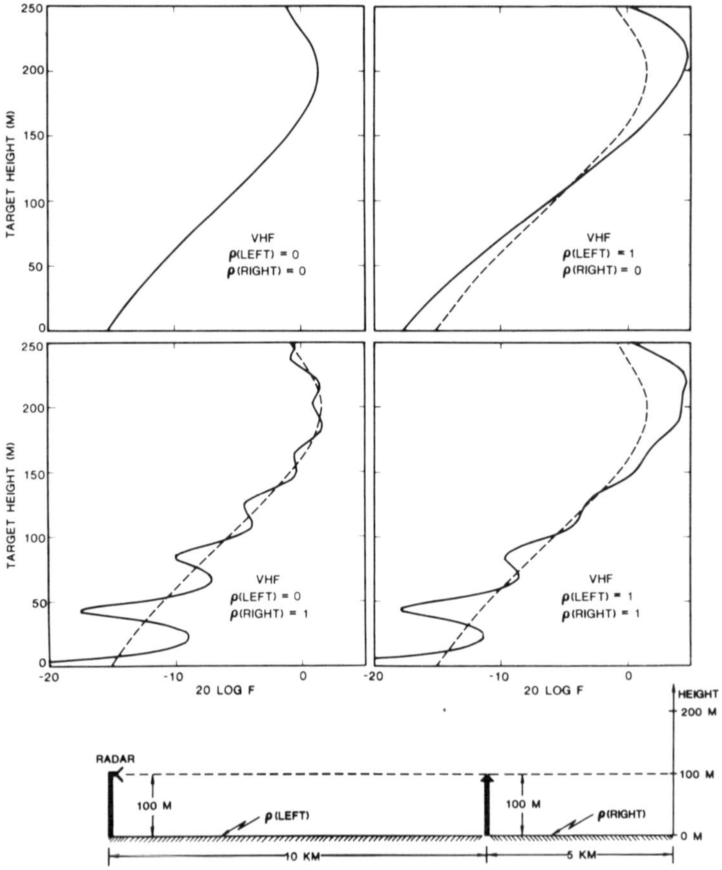

Fig. 6.3 Propagation over a knife-edge on a plane at a frequency of 1.3 GHz. The propagation geometry is shown in the bottom diagram; the four graphs show the cases of perfect reflection or no reflection on each side of the knife-edge.

relative to free space (20 log F) is calculated for a range of target heights 0 to 250 m, (*a*) with $\rho = 0$ on both sides of the mask, (*b*) for reflection on the left side only, (*c*) for reflections on the right side only, and (*d*) for reflections on both sides. These four cases are shown for an L-band radar frequency (1.3 GHz) in Figure 6.3. When ρ (left) and ρ (right) are both zero, we get the familiar Fresnel diffraction curve. When reflection occurs on the left only, we get a hybrid situation with lobe structure above 200 m where the direct and reflected waves interfere, and diffraction of both waves into the shadow zone with little change in their relative phase. When reflec-

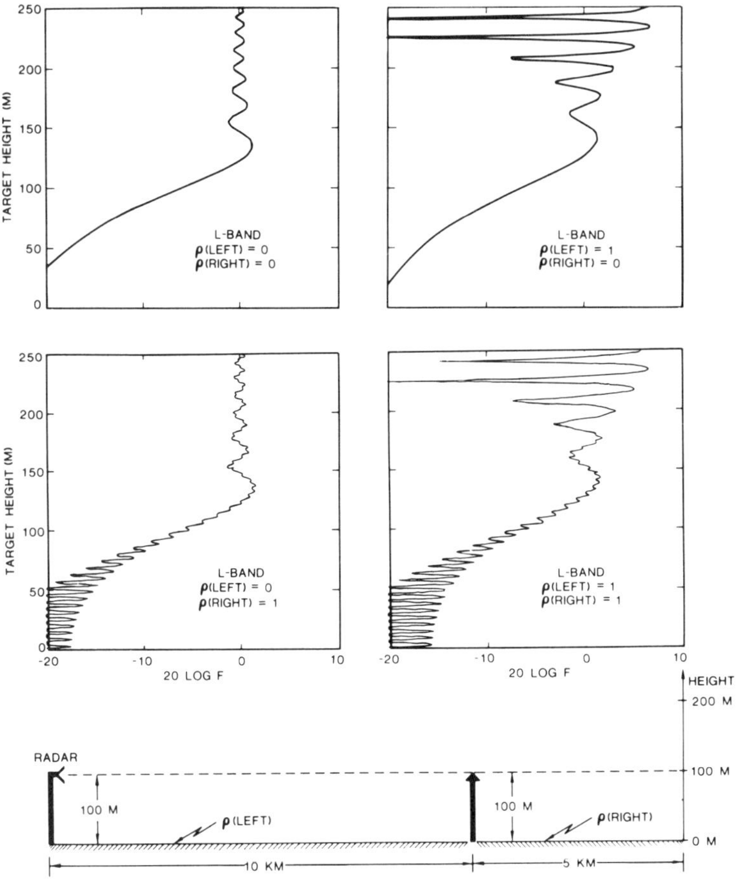

Fig. 6.4 Propagation over a knife-edge in a plane at a frequency of 170 HMz. The propagation geometry is identical with that of Fig. 6.3. The dotted line shows the curve with no reflection for comparison.

tion occurs on the right only, we find interference lobing between the direct and reflected wave with increasingly deep nulls as the target moves into the shadow zone. We can think of the knife-edge as the source of the diffracted wave in the shadow zone, and the lobe width is determined by the height of the knife-edge above the reflecting plane. When reflection occurs on both sides of the mask, the propagation combines the characteristics of the two cases of single reflection.

A similar computation made for a VHF frequency (170 MHz) is shown in Figure 6.4; the geometrical arrangement is identical with that of Figure 6.3. The target-height scale for this radar frequency goes only up to 250 m, which is not high enough to show the lobe pattern that is produced by reflections in the foreground of the radar. With reflections only on the left side of the mask, the signal strength is increased for heights from 150 to 250 m and decreased for heights below 100 m as compared with the diffraction-only case. With reflections on the target side of the mask, one finds lobes in the shadow zone which result from interference between waves from the virtual source at the knife-edge as in the L-band case. The dotted curves in Figure 6.4 represent the diffraction-only case for comparison.

6.4 Propagation Over a Spherical Earth

The problem of radio propagation over a smooth spherical earth is discussed in detail by Kerr (B 1951), by Fock (B 1965), by Domb and Pryce (J 1946), and by many others. We describe here the solutions to this classical problem; the references cited above should be consulted for a complete discussion of the problem. The radius of the sphere is, of course, taken to be the effective earth radius, i.e., radius of curvature of the terrain profile corrected for atmospheric refraction. For a radar antenna at height h_1, the range r_1 to the horizon is given by $\sqrt{2h_1\,K\,R_e}$, and looking from the target toward the radar, the range r_2 to the horizon is $\sqrt{2h_2\,K\,R_e}$, where h_2 is the target height. Hence, the range r at which the line-of-sight from radar to target is just tangent to the earth (i.e., the lowest line-of-sight) is given by:

$$r_1 + r_2 = \sqrt{2h_1\,KR_e} + \sqrt{2h_2\,KR_e}\ .$$

These results may be derived from simple geometrical considerations.

The solution to the problem is usually considered in three separate regions. Figure 6.5 shows how these regions are defined in terms of the lowest line-of-sight. In the *interference region* the propagation is calculated by geometrical optics and includes interference of the direct and reflected waves. The path difference is computed from geometrical considerations, and the amplitude of the reflected wave is corrected for the increased divergence of the rays produced by reflection from a convex surface. The interference region includes almost all target locations where an unbroken line of sight exists between radar antenna and target except for locations very near the horizon. We have programmed the interference-region computations of the pattern-propagation function F; a description of this program is contained in Appendix B. As an example of this calculation, we show in Figure 6.6 the distributions of values of F, at various ranges, for a transmitting antenna at a height of 30 m above a smooth spherical earth (reflection coefficient = 0.9 and K = 4/3) calculated for wavelength λ = 23 cm. We see the characteristic lobe structure represented in this plot. Note that the nulls and peaks line up along straight lines in Figure 6.6. The figure shows values of F, which vary between 1.9 (peaks) and 0.1 (nulls). The pattern of the transmitting antenna is assumed to effectively cut off the signal in the upper left corner.

To describe the propagation in the intermediate and diffraction regions, we must solve the much more difficult problem of the diffraction of electromagnetic waves by a smooth dielectric sphere. The general solution to this problem can be expressed in terms of an infinite series whose terms involve Airy functions of complex argument. A complete discussion of the problem may be found in Fock (B 1965), Chapters 12 and 13. Before the use of digital computers, the numerical evaluation of this series was in practice restricted to a region in which the series could be accurately represented by the

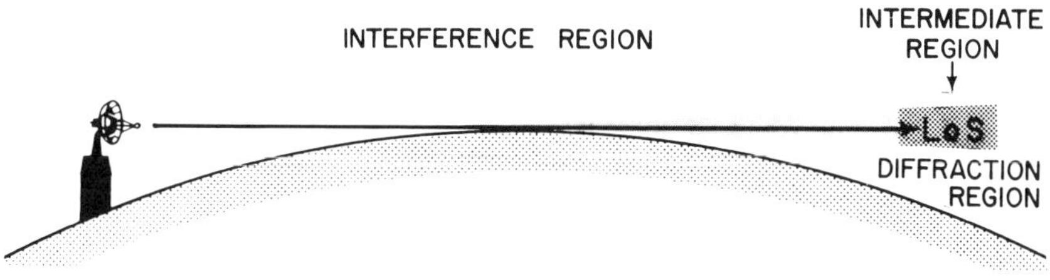

Fig. 6.5 Propagation regions over a smooth spherical earth.

first term. This region, called the *diffraction region*, lies in the shadow well below the lowest line of sight. Between the interference region and the diffraction region lies the so-called *intermediate region*, where Kerr (B 1951), p. 129, proposed that the propagation be determined by a "bold interpolation" between the interference and diffraction regions when no numerical solution was available. Now it is possible to evaluate a large number of terms in the series solution, and it is no longer necessary to distinguish a separate intermediate region. Appendix C gives a computer program for calculating the pattern-propagation factor F throughout the region in which the series solution applies. This program evaluates each term in the series by numerically integrating differential equations given in Fock (B 1965), and by terminating the series when the contribution of the last term becomes negligible.

For low-flying aircraft, we now consider propagation near the horizon where both the geometrical-optics and the sphere-diffraction

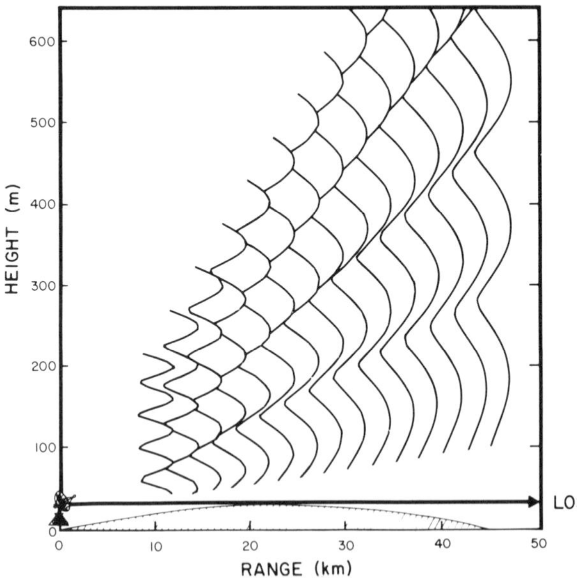

Fig. 6.6 The space distribution of the pattern propagation function for the interference region over a spherical earth. Parameters are given in the text.

solutions are necessary. Figure 6.7 shows plots of F *vs.* altitude for a target range of 30 km and an antenna height of 30 m. For VHF, L-band, and X-band frequencies, this figure shows curves representing these two solutions to the propagation problem. Note that the two solutions in Figure 6.7 come into excellent agreement near the maximum of the lowest lobe. For the X-band wavelength we continued the diffraction computation through the second lobe above the horizon in Figure 6.7. The sphere-diffraction program evaluated 27 terms in the series expansion to calculate the loss at a height of 158 m, a computation which required 10 s on a large-scale digital computer. As the target height drops below the peak of the lowest lobe, the geometrical-optics solution deviates increasingly from the diffraction solution. We have plotted the geometrical-optics curve with a dotted line in Figure 6.7. Clearly there is no sharp boundary between the valid regions of the two solutions, but because the geometrical-optics solution is much easier to compute, it seems appropriate to use this solution when the path difference between the direct and reflected rays is greater than about $\lambda/2$.

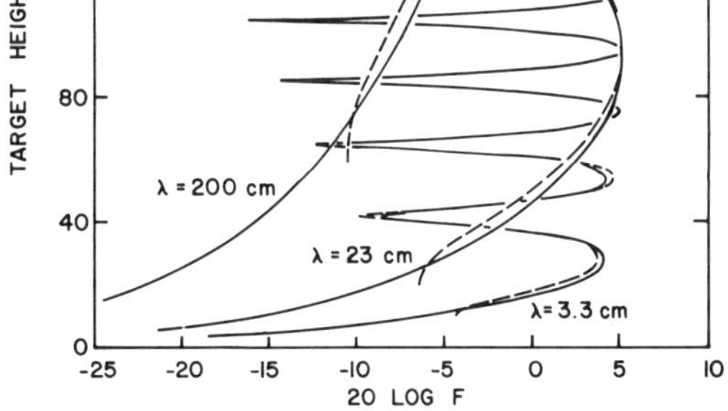

Fig. 6.7 The pattern-propagation factor *vs.* target altitude for a smooth spherical earth computed by deometrical optics and by sphere diffraction. The geometrical-optics curves are dashed where they differ appreciably from the sphere-diffraction curves. The lowest line-of-sight corresponds to a target height of 3 m.

Figure 6.7 shows clearly that over a smooth spherical earth the VHF wavelength suffers a greater loss than L-band wavelengths at altitudes below 150 m, and L-band wavelengths suffer greater loss than X-band below 40 m. For the range of altitudes shown in Figure 6.7, almost the entire curve for VHF wavelengths lies below the interference region, whereas at X-band most of the curve lies in the interference region. The null depths for the X-band curve differ in depth because of the divergence factor; a reflection coefficient of one is assumed for the geometrical-optics calculation.

Because the sphere-diffraction model represents the earth as a perfectly smooth sphere, it is not immediately clear where this model will break down as the terrain becomes increasingly rough. On the other hand, the geometrical-optics model easily allows us to take into account certain kinds of rough terrain by introducing reflection coefficients that are significantly smaller than one. But we should remember that the grazing angle becomes very small at the altitude where the transition occurs between these two models, and as noted in Section 4.3, the reflection coefficient of rough terrain tends to approach one as the grazing angle approaches zero. This behavior is a property of the Gaussian model and is a general consequence of the Rayleigh roughness criterion. Hence we may expect that the combined geometrical-optics and earth-diffraction model can predict propagation over terrain whose roughness is large compared with wavelength provided that there are no isolated obstacles along the path. However, the dependence of reflection coefficient on grazing angle must be taken into account in the geometrical-optics calculation.

7. SUMMARY

Low-altitude propagation at the frequencies considered here is controlled by atmospheric refraction and by diffraction and reflection from the terrain over which the waves travel. Where these phenomena can be taken into account quantitatively, we can predict accurately the propagation loss as a function of frequency over specific terrain. The underlying physical principles are understood, but a number of difficulties remain before the low-altitude propagation problem can be considered solved.

Refraction plays an important role in the propagation of radar waves over large bodies of water, particularly at low altitudes over the ocean and near the coast. Electromagnetic waves may be trapped in surface layers of air containing a concentration of water vapor and an inversion of the temperature gradient. Greatly extended radar ranges in such ducts are encountered frequently in maritime environments. On the other hand, over continental land masses ducting at ground level is seldom encountered except over areas of desert or snow cover when nocturnal temperature inversions caused by radiational cooling significantly affect propagation. But usually the atmosphere near the surface over land has a refractivity that decreases nearly linearly with increasing altitude. This situation, encountered roughtly 80% of the time, can be modeled by increasing the effective earth radius by a factor which varies between one and two, a typical value being 4/3. Taking refraction into account is straightforward in this case, and the resulting effect on radar coverage is comparatively small and independent of frequency.

Diffraction phenomena, on the other hand, are of major inportance in low-altitude propagation, and we must distinguish between two cases. In the first case, specific irregularities in the terrain profile produce diffraction; these terrain features can be represented by knife-edges, or in some cases by cylinders. In the second case diffraction is produced by the curvature of the smooth spherical earth, and this case applies to propagation over oceans or plains. In both cases the propagation is strongly dependent on radar frequency; the boundary between the illuminated region and the shadow becomes increasingly sharp as the frequency increases. At lower frequencies diffraction by hills or ridges alters the field strength in a significant region above and below the lowest line of sight. Diffraction reduces the signal above the mask and directs some of this energy into the shadow region below the mask. Consequently, at lower frequencies more than one diffracting feature on the terrain profile may contribute to the propagation loss. This presents a problem that has yet to be completely solved: we have only the Deygout approximation for more than two successive knife-edges. Furthermore, we do not know the extent to which a solution to the problem of propagation over a smooth spherical earth is applicable when the terrain becomes rough. These problems require thorough investigation. For example, we should be able to specify the conditions under which the multiple-diffraction case goes over to the spherical earth case as the terrain roughness decreases.

When an unbroken line-of-sight exists between radar and target, the coherent reflection from terrain must be taken into account for propagation paths over water and plains, particularly at lower frequencies. The lobe structure produced by ground reflection includes a series of nulls beginning with a null at the horizon, and the depths of these nulls depend on the reflection coefficient. Thus coherent reflection from terrain can alter low-altitude propagation significantly depending on the value of the reflection coefficient. If the terrain is smooth, the conductivity and dielectric constant of the surface determine the reflection coefficient, but for real terrain the roughness must be taken into account. For barren terrain the Gaussian terrain model may be used to estimate the effect of terrain irregularity, but this model predicts less reflection than that observed at short wavelengths and very low grazing angles. Furthermore, the reflection properties of terrain with vegetative cover are almost unknown from grazing angles below 1°. Thus extensive measurements over barren and vegetated terrain should be made to guide the development of a valid model for reflection effects at low grazing angles.

The propagation problems that have been solved include the following: (*a*) propagation over a plane with arbitrary reflectivity, (*b*) propagation over a sphere with arbitrary reflectivity in the region above the horizon where the direct and reflected waves interfere, (*c*) propagation over a smooth spherical earth with arbitrary dielectric properties in the intermediate and diffraction regions, (*d*) propagation over a knife-edge or cylinder on a reflecting plane, and (*e*)propagation over successive knife-edges treated approximately by the method of Deygout (J 1966). In many cases these models are adequate for predicting propagation loss provided the reflection coefficients are known.

APPENDIX A
Four-Ray Propagation Model

This model calculates the pattern propagation factor F for radio propagation over flat terrain on which a knife-edge obstruction lies perpendicular to the direction of propagation. The geometrical arrangement is shown in Figure A-1. The program computes F for a sequence of values of Z_2, the height of R above ground.

Input

Geometrical distances in meters Z_1, d_1, d_2, f; and the initial value, final value, and step for Z_2.

Wavelength λ in meters

Polarization: Horizontal (H) or Vertical (V)

Ground properties on left and right side of the mask: relative dielectric constant ε_r, conductivity σ (in mhos/m)

(Note: The reflection coefficient on either or both sides may be taken as zero if $\varepsilon = \sigma + 888$.)

In case the surface is rough, the reflection coefficient may be altered by factors REF L and REF R on the left and right sides of the mask.

Output

Table with the following columns: Height of R: Z2 (in meters), F (pattern propagation factor), and 20 log F.

For Rays 1 and 2: v_1, $|E_1|$, ρ_1, ϕ_1, and v_2, $|E_2|$, p_2, ϕ_2
For Rays 4: v_4 and $|E_4|$

Reflections on the left and right sides are taken into account by including the images of R and T, which are denoted by S and U, respectively. The four possible rays to be considered are TR, UR, TS and US, which are designated as Rays 1, 2, 3, and 4, respectively. The reflection coefficients are designated by the subscripts 1 and 2 for the left and right sides, respectively. Each reflection coefficient is given by a complex number calculated from the Fresnel reflection formulas in a subroutine, FRESNL, and has the form

$$\Gamma = \rho^{-j\Phi}$$

where the magnitude ρ and the phase lag Φ depend on the grazing angle, the polarization, and the electrical constants, ε_r and σ, of the ground. The Fresnel integrals C(v), and S(v), are evaluated in a subroutine, DCS, where $v = \sqrt{2}c$ with c equal to the ray clearance over the knife-edge. (When the ray intersects the mask, c is taken as negative.) The pattern propagation factor is equal to |F|, where

$$F = \sum_{k=1}^{4} E_k \exp(j\psi)$$

and

$$E_1 = A_1 \qquad \psi_1 = \beta_1$$

$$E_2 = \Gamma_1 A_2 \qquad \psi_2 = \beta_2 + \frac{2\pi}{\lambda}(R_1 - R_2)$$

$$E_3 = \Gamma_2 A_3 \qquad \psi_3 = \beta_3 + \frac{2\pi}{\lambda}(R_3 - R_1)$$

$$E_4 = \Gamma_1 \Gamma_2 A_4 \qquad \psi_4 = \beta_4 + \frac{2\pi}{\lambda}(R_4 - R_1)$$

$$A = \frac{1}{\sqrt{2}} \left[(C(v) + 1/2)^2 + (S(v) + 1/2)^2 \right]^{1/2}$$

$$\beta = \tan^{-1}\left(\frac{S + 1/2}{C + 1/2}\right) - \pi/4 \text{ if } (C + 1/2) \geq 0$$

$$\beta = \pi + \tan^{-1}\left(\frac{S + 1/2}{C + 1/2}\right) - \pi/4 \text{ if } (C + 1/2) < 0$$

This computation was programmed in IBM Fortran IV ANSI by Gerald McCaffrey; a copy of the listings for this program as well as a sample of the output follow this description. The sample computation appears plotted as Figure 6.4, and all the input parameters are listed on the output.

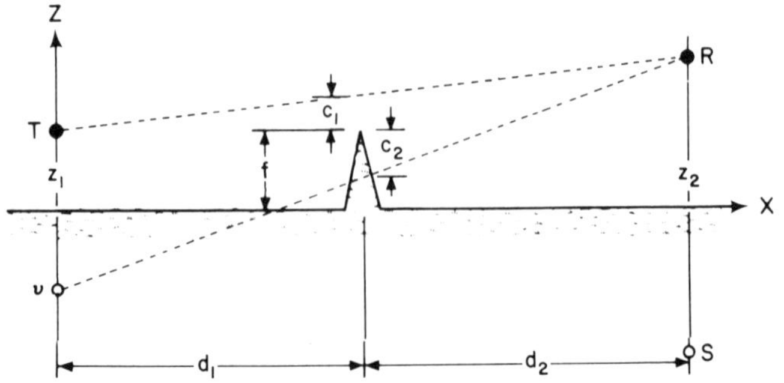

Fig. A-1 Geometrical arrangement in the four-ray model.

PROGRAM LISTINGS: FOUR-RAY PROPAGATION MODEL

PROGRAMMED BY GERALD McCAFFREY

```
C
C     VARIABLES THAT START WITH THE LETTER C ARE COMPLEX
C
      COMPLEX CT1,CT2,CF,CE1,CE2,CE3,CE4,CTH,CTV
      REAL*8 R1,R2,R3,R4,VC1,VC2,VC3,VC4,VS1,VS2,VS3,VS4,A,B,V1,V2,V3,V4
      REAL*8 VTERM
      REAL*4 LAMDA
      DIMENSION POLAR(2)
      DATA POLAR/'H   ','V   '/
      PI=3.14159
      RAD=57.29578
C
C     GET SQRT(2.0)
C
      SQRT2=SQRT(2.0)
C
C     READ GEOMETRIC PARAMETERS
C
      WRITE(6,1)
1     FORMAT(' INPUT Z1,D1,D2,F,Z2S,Z2E,DZ2')
      READ(5,*) Z1,D1,D2,F,Z2S,Z2E,DZ2
C
C     READ IN DIALECTIC CONSTANTS AND FRESNEL FACTORS (LEFT AND RIGHT)
C
      WRITE(6,4)
4     FORMAT(' INPUT PSIL,SL,PSIR,SR,REFL,REFR')
      READ(5,*) EL,SL,ER,SR,REFL,REFR
C
C     READ IN POLARIZATION
C
      WRITE(6,2)
2     FORMAT(' INPUT POLARIZATION.. H=HORIZONTAL,V=VERTICAL')
      READ(5,3) POLR
3     FORMAT(A4)
      IF(POLR.EQ.POLAR(1)) IPOL=1
      IF(POLR.EQ.POLAR(2)) IPOL=2
C
C     INITIALIZE Z2
C
      Z2=Z2S-DZ2
C
C     SET WAVELENGTH
C
```

```
      LAMDA=.302
C
C      WRITE OUT THE INPUTS AND SET UP HEADER FOR THE PRINTOUT
C
      WRITE(1,10) Z1,D1,D2,F,Z2S,Z2E,DZ2,LAMDA,POLR,EL,ER,SL,SR
      WRITE(1,9)
9     FORMAT('    ')
      WRITE(6,10) Z1,D1,D2,F,Z2S,Z2E,DZ2,LAMDA,POLR,EL,ER,SL,SR
     1,REFL,REFR
10    FORMAT(' ANNT HT',F6.2,' D1',F6.0,' D2',F6.0,' F',F6.1,15X,
     1' TARGET HT FROM',F5.0,' TO',F5.0,' BY',F5.0,/' WAVELENGTH',F6.3
     3' POLARIZATION ',A4,' EPSILON L',F8.3,' EPSILON R',F8.3,' SIG L'
     4F8.3,' SIG R',F8.3/,' REFL',F10.2,' REFR',F10.2)
      WRITE(1,11)
11    FORMAT(2X,'TARGET HT',2X,' F ',1X,'20LOG(F)',1X,' V(1)',1X,
     1' MAG(E1)',2X,' RHO(1)',1X,' PHI(1)',1X,' V(2)',1X,' MAG(E2)',3X
     2' V(3)',1X,' MAG(E3)',1X,' RHO(2)',1X,' PHI(2)',3X,' V(4)',1X,
     3' MAG(E4)')
C
C      CALCULATE HO
C
      HO1=LAMDA*D1*D2
      HO2=D1+D2
      HO=SQRT(HO1/HO2)
C
C      LOOP ON Z2 FROM Z2S TO Z2E BY DZ2
C
100   Z2=Z2+DZ2
      IF(Z2.GT.Z2E) GO TO 1000
C
C      CALCULATE R1,R2,R3,AND R4
C
      R1=DSQRT((Z2*1.0D0-Z1)**2+(D1+D2)**2)
      R2=DSQRT((-Z2*1.0D0-Z1)**2+(D1+D2)**2)
      R3=R2
      R4=R1
C
C      CALCULAT V1,V2,V3,AND V4
C
      VTERM=((Z2*1.0D0-Z1)*D1)/(D1+D2)
      V1=SQRT2/HO*(Z1+VTERM-F)
C
      VTERM=((Z2*1.0D0+Z1)*D1)/(D1+D2)
      V2=SQRT2/HO*(-Z1+VTERM-F)
C
      VTERM=((-Z1*1.0D0-Z2)*D1)/(D1+D2)
      V3=SQRT2/HO*(Z1+VTERM-F)
C
      VTERM=((Z1*1.0D0-Z2)*D1)/(D1+D2)
      V4=SQRT2/HO*(-Z1+VTERM-F)
C
C      CALCULATE THE GRAZING ANGLES
```

```
C
C       LEFT GRAZING ANGLE
C
      IF(V2.GE.0) GO TO 200
C
C       RAY INTERSECTS MASK
C
      PSIL=DATAN((Z1+F)/D1*1.0D0)
      GO TO 201
C
C       RAY CLEAR OF MASK
C
200   PSIL=DATAN((Z2+Z1)/(D1+D2*1.0D0))
C
C       RIGHT GRAZING ANGLE
C
201   IF(V3.GE.0) GO TO 205
C
C       RAY INTERSECTS MASK
C
      PSIR=DATAN((Z2+F)/D2*1.0D0)
      GO TO 206
C
C       RAY CLEAR OF MASK
C
205   PSIR=DATAN((Z2+Z1)/(D1+D2*1.0D0))
C
C       CALL THE COMPLEX REFLECTION COEFFICIENT SUBROUTINE
C
C       FOR THE LEFT
C
206   CALL FRESNL(EL,LAMDA,SL,PSIL,CTH,CTV)
C
C       MULTIPLY BY FACTORS
C
      CTH=CTH*REFL
      CTV=CTV*REFL
C
C       SET CT1
C
      IF(IPOL.EQ.1) CT1=CTH
      IF(IPOL.EQ.2) CT1=CTV
C
C       888 SHOWS NO REFLECTION ON THE LEFT
C
      IF(EL.EQ.888.AND.SL.EQ.888) CT1=CMPLX(0.0,0.0)
C
C       FOR THE RIGHT
C
      CALL FRESNL(ER,LAMDA,SR,PSIR,CTH,CTV)
C
C       MULTIPLY BY FACTORS
```

```
C
      CTH=CTH*REFR
      CTV=CTV*REFR
C
C      SET CT2
C
      IF(IPOL.EQ.1) CT2=CTH
      IF(IPOL.EQ.2) CT2=CTV
C
C      888 SHOWS NO REFLECTION ON THE RIGHT
C
      IF(ER.EQ.888.AND.SR.EQ.888) CT2=CMPLX(0.0,0.0)
C
C      GET THE REAL AND IMAGINARY PARTS FOR VERT AND HORZ
C
      P1R=REAL(CT1)
      P1I=AIMAG(CT1)
      P2R=REAL(CT2)
      P2I=AIMAG(CT2)
C
C      GET THE PHASE LAG IN DEGREES
C
      THETA1=0
      THETA2=0
      IF(P1I.NE.0.AND.P1R.NE.0)THETA1=ATAN2(P1I,P1R)*(-RAD)
      IF(P2I.NE.0.AND.P2R.NE.0)THETA2=ATAN2(P2I,P2R)*(-RAD)
C
C      GET THE REFLECTION COEF
C
      PR1=CABS(CT1)
      PR2=CABS(CT2)
C
C      CALCULATE COMPLEX F (CF)
C       BY FIRST CALCULATING THE COMPLEX PARTS CE1,CE2,CE3,AND CE4
C
C      CALCULATE CE1 USING V1
C
      CALL DCS(VC1,VS1,V1)
      A=1/SQRT2*DSQRT((VC1+0.5)**2+(VS1+0.5)**2)
      IF(VC1+0.5.GE.0) B=DATAN((VS1+0.5)/(VC1+0.5))-PI/4
      IF(VC1+0.5.LT.0) B=PI+DATAN((VS1+0.5)/(VC1+0.5))-PI/4
      A1=A*DSIN(B)
      B1=A*DCOS(B)
      CE1=CMPLX(B1,A1)
C
C      CALCULATE CE2 USING V2
C
      CALL DCS(VC2,VS2,V2)
      A=1/SQRT2*DSQRT((VC2+0.5)**2+(VS2+0.5)**2)
      IF(VC2+0.5.GE.0) B=DATAN((VS2+0.5)/(VC2+0.5))-PI/4
      IF(VC2+0.5.LT.0) B=PI+DATAN((VS2+0.5)/(VC2+0.5))-PI/4
      B=B+(2*PI)/LAMDA*(R2-R1)
```

```
      A1=A*DSIN(B)
      B1=A*DCOS(B)
      CE2=CMPLX(B1,A1)*CT1
C
C      CALCULATE CE3 USING V3
C
      CALL DCS(VC3,VS3,V3)
      A=1/SQRT2*DSQRT((VC3+0.5)**2+(VS3+0.5)**2)
      IF(VC3+0.5.GE.0) B=DATAN((VS3+0.5)/(VC3+0.5))-PI/4
      IF(VC3+0.5.LT.0) B=PI+DATAN((VS3+0.5)/(VC3+0.5))-PI/4
      B=B+(2*PI)/LAMDA*(R3-R1)
      A1=A*DSIN(B)
      B1=A*DCOS(B)
      CE3=CMPLX(B1,A1)*CT2
C
C      CALCULATE CE4 USING V4
C
      CALL DCS(VC4,VS4,V4)
      A=1/SQRT2*DSQRT((VC4+0.5)**2+(VS4+0.5)**2)
      IF(VC4+0.5.GE.0) B=DATAN((VS4+0.5)/(VC4+0.5))-PI/4
      IF(VC4+0.5.LT.0) B=PI+DATAN((VS4+0.5)/(VC4+0.5))-PI/4
      A1=A*DSIN(B)
      B1=A*DCOS(B)
      CE4=CT1*CT2*CMPLX(B1,A1)
C
C      NOW ADD TO GET CF
C
      CF=CE1+CE2+CE3+CE4
C
C      WRITE OUT ANSWERS
C
      FO=CABS(CF)
      FLOG=20*ALOG10(FO)
      E1MAG=CABS(CE1)
      E2MAG=CABS(CE2)
      E3MAG=CABS(CE3)
      E4MAG=CABS(CE4)
      WRITE(1,99) Z2,FO,FLOG,V1,E1MAG,PR1,THETA1,V2,E2MAG,V3,E3MAG,
     1PR2,THETA2,V4,E4MAG
99    FORMAT(F8.1,4(2F7.2,F10.3),2F8.2)
      GO TO 100
1000  RETURN
      END
```

SUBROUTINE FOR COMPLEX REFLECTION COEFFICIENTS

```
      SUBROUTINE FRESNL(El,WAVE,CONDUC,ANG,CTH,CTV)
      COMPLEX CAK,CTV,CTV1,CTV2,CTH,CTH1,CTH2
C
C       El.......THE DIELECTIC CONSTATNT (FROM 0 TO 100)
C       LAMDA....THE WAVELENGTH IN METERS
C       CONDUC...THE CONDUCTIVITY IN MHOS/METER
C       ANG......THE ANGLE IN RADIANS
C
C
C      CALCULATE THE COMPLEX CONSTANT
C
      AKI=-60*WAVE*CONDUC
      CAK=CMPLX(El,AKI)
C
C      CALCULATE THE VERTICAL POLARIZATION
C
      CTV1=CAK*SIN(ANG)-CSQRT(CAK-COS(ANG)**2)
      CTV2=CAK*SIN(ANG)+CSQRT(CAK-COS(ANG)**2)
      CTV=CTV1/CTV2
C
C      CALCULATE THE HORIZONTAL
C
      CTH1=SIN(ANG)-CSQRT(CAK-COS(ANG)**2)
      CTH2=SIN(ANG)+CSQRT(CAK-COS(ANG)**2)
      CTH=CTH1/CTH2
1000  RETURN
      END
```

SUBROUTINE TO EVALUATE THE FRESNEL INTEGRALS

```
      SUBROUTINE DCS (C,S,X)
      IMPLICIT REAL*8 (A-H,O-Z)
      DIMENSION CC(12),DD(12),AA(12),BB(12)
      U=X
      PIE2=1.5707963268D0
      X=PIE2*X*X
      Z=DABS(X)
      IF(Z.NE. 0.D0) GO TO 1000
      C=0.D0
      S=0.D0
      X=U
      RETURN
1000  CONTINUE
      IF(Z-4.0D0)3,3,4
 3    C=DCOS(Z)
      S=DSIN(Z)
      Z=Z/4.0D0
      DZ=DSQRT(Z)
```

```
      AA(1)=0.159576914D+01
      AA(2)=-0.1702D-05
      AA(3)=-0.6808568854D+01
      AA(4)=-0.576361D-03
      AA(5)=0.6920691902D+01
      AA(6)=-0.16898657D-01
      AA(7)=-0.305048566D+01
      AA(8)=-0.75752419D-01
      AA(9)=0.850663781D0
      AA(10)=-0.25639041D-01
      AA(11)=-0.15023096D0
      AA(12)=0.34404779D-01
      BB(1)=-0.33D-07
      BB(2)=0.4255387524D+01
      BB(3)=-0.9281D-04
      BB(4)=-0.77800204D+01
      BB(5)=-0.9520895D-02
      BB(6)=0.5075161298D+01
      BB(7)=-0.138341947D0
      BB(8)=-0.1363729124D+01
      BB(9)=-0.403349276D0
      BB(10)=0.702222016D0
      BB(11)=-0.216195929D0
      BB(12)=0.19547031D-01
      ASUM=AA(1)
      BSUM=BB(1)
      DO 40 J=2,12
      ASUM=ASUM+AA(J)*Z**(J-1)
      BSUM=BSUM+BB(J)*Z**(J-1)
40     CONTINUE
      FC=DZ*(S*BSUM+C*ASUM)
      FS=DZ*(-C*BSUM+S*ASUM)
      C=FC
      S=FS
      GO TO 5
4     D=DCOS(Z)
      S=DSIN(Z)
      Z=4.0D0/Z
      CC(1)=0.0D0
      CC(2)=-0.24933957D-01
      CC(3)=0.3936D-05
      CC(4)=0.5770956D-02
      CC(5)=0.689892D-03
      CC(6)=-0.9497136D-02
      CC(7)=0.11948809D-01
      CC(8)=-0.6748873D-02
      CC(9)=0.24642D-03
      CC(10)=0.2102967D-02
      CC(11)=-0.121793D-02
      CC(12)=0.233939D-03
      DD(1)=0.19947114D0
      DD(2)=0.23D-07
      DD(3)=-0.9351341D-02
      DD(4)=0.23006D-04
      DD(5)=0.485146D-02
      DD(6)=0.1903218D-02
      DD(7)=-0.17122914D-01
      DD(8)=0.29064067D-01
      DD(9)=-0.27928955D-01
      DD(10)=0.16497308D-01
      DD(11)=-0.5598515D-02
      DD(12)=0.838386D-03
      DSUM=DD(1)
      CSUM=CC(1)
      DO 30 J=2,12
      CSUM=CSUM+CC(J)*Z**(J-1)
30     DSUM=DSUM+DD(J)*Z**(J-1)
      Z=DSQRT(Z)
      C=0.5D0+Z*(D*CSUM+S*DSUM)
      S=0.5D0+Z*(S*CSUM-D*DSUM)
5     X=U
      IF(U.GT.0.D0) GO TO 6
      C=-C
      S=-S
6     RETURN
      END
```

SAMPLE OUTPUT: FOUR-RAY PROPAGATION MODEL

ANNT HT100.00 D110000. D2 5000. F 100.0 TARGET HT FROM 0. TO 250. BY 10.
WAVELENGTH 1.760 POLARIZATION H EPSILON L 15.000 EPSILON R 15.000 SIG L 0.005 SIG R 0.005
REFL

TARGET HT	F	20LOG(F)	V(1)	MAG(E1)	RHO(1)	PHI(1)	V(2)	MAG(E2)	V(3)	MAG(E3)	RHO(2)	PHI(2)	V(4)	MAG(E4)
0.0	0.00	-56.42	-1.231	0.17	0.99	-179.988	-2.46	0.09	-1.231	0.17	0.99	-179.988	-2.46	0.09
10.0	0.19	-14.50	-1.108	0.19	0.99	-179.988	-2.34	0.09	-1.354	0.16	0.99	-179.987	-2.58	0.08
20.0	0.29	-10.78	-0.985	0.21	0.99	-179.988	-2.22	0.10	-1.477	0.15	0.99	-179.986	-2.71	0.08
30.0	0.26	-11.77	-0.862	0.23	0.99	-179.988	-2.09	0.11	-1.600	0.13	0.99	-179.985	-2.83	0.08
40.0	0.14	-16.86	-0.739	0.25	0.99	-179.988	-1.97	0.11	-1.723	0.13	0.99	-179.984	-2.95	0.07
50.0	0.21	-13.67	-0.615	0.28	0.99	-179.988	-1.85	0.12	-1.846	0.12	0.98	-179.983	-3.08	0.07
60.0	0.35	-9.02	-0.492	0.31	0.99	-179.988	-1.72	0.13	-1.969	0.11	0.98	-179.982	-3.20	0.07
70.0	0.39	-8.08	-0.369	0.35	0.99	-179.988	-1.60	0.14	-2.093	0.10	0.98	-179.980	-3.32	0.07
80.0	0.34	-9.34	-0.246	0.39	0.99	-179.988	-1.48	0.15	-2.216	0.10	0.98	-179.979	-3.45	0.06
90.0	0.38	-8.39	-0.123	0.44	0.99	-179.988	-1.35	0.16	-2.339	0.09	0.98	-179.978	-3.57	0.06
100.0	0.55	-5.22	0.0	0.50	0.99	-179.988	-1.23	0.17	-2.462	0.09	0.98	-179.977	-3.69	0.06
110.0	0.66	-3.65	0.123	0.57	0.99	-179.988	-1.11	0.19	-2.585	0.08	0.98	-179.976	-3.82	0.06
120.0	0.67	-3.46	0.246	0.64	0.99	-179.988	-0.98	0.20	-2.708	0.08	0.98	-179.975	-3.94	0.06
130.0	0.75	-2.48	0.369	0.72	0.99	-179.988	-0.86	0.22	-2.831	0.08	0.98	-179.973	-4.06	0.05
140.0	0.96	-0.39	0.492	0.80	0.99	-179.988	-0.74	0.25	-2.954	0.07	0.97	-179.972	-4.19	0.05
150.0	1.11	0.94	0.615	0.89	0.99	-179.988	-0.62	0.27	-3.077	0.07	0.97	-179.971	-4.31	0.05
160.0	1.18	1.44	0.739	0.97	0.99	-179.988	-0.49	0.31	-3.200	0.07	0.97	-179.970	-4.43	0.05
170.0	1.31	2.33	0.862	1.05	0.99	-179.988	-0.37	0.34	-3.323	0.07	0.97	-179.969	-4.55	0.05
180.0	1.51	3.60	0.985	1.12	0.99	-179.988	-0.25	0.39	-3.447	0.06	0.97	-179.968	-4.68	0.05
190.0	1.62	4.20	1.108	1.16	0.99	-179.988	-0.12	0.44	-3.570	0.06	0.97	-179.967	-4.80	0.04
200.0	1.62	4.21	1.231	1.17	0.99	-179.988	0.0	0.49	-3.693	0.06	0.97	-179.965	-4.92	0.04
210.0	1.66	4.41	1.354	1.15	0.99	-179.988	0.12	0.56	-3.816	0.06	0.97	-179.964	-5.05	0.04
220.0	1.65	4.37	1.477	1.09	0.99	-179.988	0.25	0.63	-3.939	0.06	0.97	-179.963	-5.17	0.04
230.0	1.45	3.25	1.600	1.01	0.99	-179.987	0.37	0.71	-4.062	0.05	0.97	-179.962	-5.29	0.04
240.0	1.15	1.24	1.723	0.93	0.99	-179.987	0.49	0.79	-4.185	0.05	0.96	-179.961	-5.42	0.04
250.0	0.84	-1.48	1.846	0.88	0.99	-179.987	0.62	0.88	-4.308	0.05	0.96	-179.960	-5.54	0.04

APPENDIX B

The Smooth Spherical-Earth Model For The Interference Region

This model implements in IBM Fortran IV ANSI the computation of propagation in the interference region over a smooth earth as described in Kerr (B 1951), pp. 112-122.

For a specified radar wavelength, transmitting antenna height, and earth reflection coefficient, the program described here computes the pattern-propagation factor for a sequence of target heights at a specified target range. Refraction is taken into account by specifying the earth-radius factor K; the true radius is taken as 6,373 km. All parameters and variables are designated by the symbols used in that reference unless specified otherwise; the input and output quantities are as follows:

Input

Target range (km) RNG

Transmitter antenna height |(m) Z1|
Earth-radius factor AK

Wavelength (cm) WAVE

Reflection coefficient

Magnitude RHO

Phase PHAZ

Target height (m)

Starting height Z2S

Ending height Z2E

Incremental step DZ2

Output

Table with the following columns:

Target height

Pattern propagation factor F

F, 20 log F, and 40 log F

Grazing angle

Distances

R_1 transmitter to specular point
R_2 target to specular point

Divergence factor

ΔR path difference between direct and reflected waves

First-Fresnel-zone size

S-MAJ semimajor axis and S-MIN semiminor axis of first Fresnel zone

Note that the output contains a number of geometrical quantities that are useful in evaluating the model predictions. For example, by knowing the path difference ΔR, one can locate the boundary of the interference region, $\Delta R = \lambda/2$. The dimensions of the first Fresnel zone are calculated approximately by constructing a plane tangent to the earth at the specular point and by calculating the dimensions of the semimajor and semiminor axes of the Fresnel ellipse on this tangent plane [see Kerr, B1951, p. 415, Eqs. (31)-(33)].

This computation was programmed by Gerald McCaffrey; a copy of the program listings and a sample output, which includes a list of input parameters, follows this description. The sample output is plotted for λ = 3.3 cm in Figure 6.7.

PROGRAM LISTINGS: SPHERICAL-EARTH MODEL, INTERFERENCE REGION

PROGRAMMED BY GERALD McCAFFREY

```
      DATA PI/3.14159/
      DATA RAD/57.2957795/
      DATA RADIUS/6373.0/
C
C     READ INPUTS, FROM INPUT FORM
C
      WRITE(6,10)
10    FORMAT(' INPUT TRAN HT,WAVELENGTH,EARTH RAD FACTOR,RHO AND PHASE')
      READ(5,*) Z1,WAVE,AK,RHO,PHAZ
      PZ1=Z1
C
C     CONVERT PHAZ TO RADIANS
C
      RPHZ=PHAZ/RAD
C
C     CALCULATE AE = THE EFFECTIVE EARTH RADIUS
C
      AE=RADIUS*AK
C
C     GET RANGE AND TARGET LOOPS SETUP
C
      WRITE(6,204)
204   FORMAT(' INPUT RANGE LOOP : RNGS,RNGE,DRNG')
      READ(5,*) RNGS,RNGE,DRNG
      WRITE(6,203)
203   FORMAT(' INPUT TARGET HEIGHT LOOP :Z2ST,Z2E,DZ2')
      READ(5,*) Z2ST,Z2E,DZ2
      Z2S=Z2ST-DZ2
C
C     LOOP ON RANGE
C
      RNG=RNGS-DRNG
200   RNG=RNG+DRNG
      IF(RNG.GT.RNGE) GO TO 1000
      IC=0
      IC1=0
C
C     LOOP ON TARGET HEIGHT
C
      Z2S=Z2ST-DZ2
 250  Z2S=Z2S+DZ2
      IF(Z2S.GT.Z2E) GO TO 200
      Z2=Z2S
      PZ2=Z2
```

```
C     CHECK THAT Z1.GE.Z2
C
      Z1=PZ1

      IF(Z1.GE.Z2) GO TO 20
C
C     IF NOT SWITCH Z1 AND Z2
C
      SAVE=Z1
      Z1=Z2
      Z2=SAVE
C
20    IC=IC+1
      IC1=IC1+1
C
C     CALCULATE R2 - DISTANCE TARGET TO REFLECTION (KM)
C
      A=AE*((Z1+Z2)/1000.0)
      B=(RNG/2.0)**2
      P=2.0/SQRT(3.0)*SQRT(A+B)
      ANG=ARCOS((Z2-Z1)*2*AE*RNG/(1000.0*P**3))
      R2=RNG/2.0+P*COS((ANG+PI)/3.0)
      R1=RNG-R2
      PR1=R1
      PR2=R2
      IF(PZ1.GE.PZ2) GO TO 25
      SAVE=PR1
      PR1=PR2
      PR2=SAVE
25    S1=R1/SQRT(2*AE*Z1/1000.0)
      S2=R2/SQRT(2*AE*Z2/1000.0)
      S=RNG/(SQRT(2*AE*Z1/1000.0)+SQRT(2*AE*Z2/1000.0))
      IF(S1.LT.1.AND.S2.LT.1.AND.S.LT.1) GO TO 30
      F=0
      F20=0
      F40=0
      X=0
      R1=0
      R2=0
      GO TO 39
30    T=SQRT(Z2/Z1)
      D1=4*S2*T*S1**2
      D2=S*(1-S1**2)*(1+T)
      D3=1+D1/D2
      D=1/SQRT(D3)
      X1=((1-S1**2)+T**2*(1-S2**2))/(1+T**2)
      X2=(Z1+Z2)/(1000.0*RNG)
      X3=X1*X2
      X=ATAN(X3)
      XDEG=X*RAD
```

```
      DR1=2*Z1*Z2/(10*RNG)
      DR2=(1-S1**2)*(1-S2**2)
      DR=DR1*DR2
      F1=1+RHO**2*D**2
      F2=2*RHO*D*COS((2*PI*DR/WAVE)+RPHZ)
      F=SQRT(F1+F2)
      F20=20*ALOG10(F)
      F40=40*ALOG10(F)
      Z1P=Z1-R1**2/(2*AE)
      Z2P=Z2-R2**2/(2*AE)
      CG=1+2*DR/WAVE
      CH=1+(Z1P+Z2P)**2/(10*WAVE*RNG)
      CI=SQRT(10*WAVE*RNG)
      CJ=1+2*Z1P*(Z1P+Z2P)/(10*WAVE*RNG)
      XO=RNG/2*CJ/CH
      IF(PZ2.GT.PZ1) XO=RNG-XO
      XC=RNG/2*SQRT(CG)/CH
      YC=CI/2*SQRT(CG/CH)
39    IF(IC.GT.1) GO TO 40
      WRITE(8,8999)
8999  FORMAT(50X,'PROP1(12/11/78 VERSION)')
      WRITE(8,9000) RNG,WAVE
9000  FORMAT('1',5X,'RANGE= ',F4.1,' KM',70X,'WAVELENGTH= ',F5.1,' CM')
      WRITE(8,9001) PZ1
9001  FORMAT(5X,'TRANSMITTER HEIGHT= ',F5.1,' M',57X,'REFLECTION COEFFIC
     1IENT')
      WRITE(8,9002) AK,RHO
9002  FORMAT(5X,'EARTH RADIUS FACTOR= ',F5.2,58X,'MAGNITUDE= ',F5.3)
      WRITE(8,9003) PHAZ
9003  FORMAT(89X,'PHASE= ',F6.1,/)
      WRITE(8,9004)
9004  FORMAT(2X,'TARGET HEIGHT',4X,'PATTERN PROPAGATION FACTOR',3X,'GRAZ
     1ING',8X,'R1',7X,'R2',7X,'DIV',5X,'DELTA-R',6X,'FIRST FRESNEL ZONE
     2')
      WRITE(8,9005)
9005  FORMAT(7X,'(M)',10X,'F',5X,'20LOG(F)',2X,'40LOG(F)',2X'ANGLE (DEG)
     1',5X,'(KM)',5X,'(KM)',5X,'FACTOR',5X,'(CM)',2X,'CENTER(KM)',1X,
     2'S-MAJ(KM)',2X,'S-MINOR(M)')
40    WRITE(8,9006) PZ2,F,F20,F40,XDEG,PR1,PR2,D,DR,XO,XC,YC
9006  FORMAT(F10.2,F13.2,F10.2,F10.2,F10.2,F13.2,F9.2,F11.3,F9.2,F9.3,F9
     1.3,F13.2)
      WRITE(1,45) RNG,PZ2,F20,F40
45    FORMAT(4F12.4)
      IF(IC1.EQ.4) WRITE(8,9007)
9007  FORMAT(' ')
      IF(IC1.EQ.4) IC1=0
      GO TO 250
1000  RETURN
      END
```

SAMPLE OUTPUT: SPHERICAL-EARTH MODEL, INTERFERENCE REGION

PROP1(12/11/78 VERSION)

1 RANGE= 30.0 KM
TRANSMITTER HEIGHT= 30.0 M
EARTH RADIUS FACTOR= 1.33

WAVELENGTH= 3.3 CM
REFLECTION COEFFICIENT
MAGNITUDE= 0.700
PHASE= 180.0

TARGET HEIGHT	PATTERN PROPAGATION FACTOR			GRAZING	R1	R2	DIV	DELTA-R	FIRST FRESNEL ZONE		
(M)	F	20LOG(F)	40LOG(F)	ANGLE (DEG)	(KM)	(KM)	FACTOR	(CM)	CENTER(KM)	S-MAJ(KM)	S-MINOR(M)
10.00	0.72	-2.82	-5.64	0.02	19.70	10.30	0.431	0.18	19.630	6.042	10.25
20.00	1.10	0.86	1.71	0.04	16.97	13.03	0.554	0.87	17.148	5.263	10.36
30.00	1.41	2.98	5.95	0.06	15.00	15.00	0.623	1.87	15.000	4.730	10.68
40.00	0.60	-4.42	-8.85	0.08	13.44	16.56	0.672	3.08	13.218	4.272	10.92
50.00	1.35	2.58	5.16	0.10	12.16	17.84	0.712	4.44	11.753	3.863	11.07
60.00	1.00	0.02	0.05	0.12	11.08	18.92	0.745	5.91	10.546	3.499	11.12
70.00	1.18	1.42	2.83	0.14	10.15	19.85	0.774	7.47	9.542	3.178	11.10
80.00	1.12	1.02	2.04	0.15	9.35	20.65	0.798	9.10	8.699	2.897	11.04
90.00	1.20	1.61	3.22	0.17	8.65	21.35	0.819	10.78	7.984	2.650	10.95
100.00	1.03	0.22	0.44	0.19	8.04	21.96	0.838	12.51	7.371	2.433	10.84
110.00	1.38	2.78	5.56	0.20	7.49	22.51	0.854	14.27	6.842	2.241	10.72
120.00	0.74	-2.61	-5.22	0.22	7.01	22.99	0.868	16.06	6.380	2.072	10.58
130.00	1.57	3.90	7.81	0.24	6.58	23.42	0.881	17.88	5.975	1.922	10.45
140.00	0.39	-8.11	-16.22	0.26	6.19	23.81	0.891	19.72	5.616	1.789	10.31
150.00	1.62	4.18	8.37	0.27	5.85	24.15	0.901	21.58	5.297	1.669	10.17
160.00	0.64	-3.86	-7.72	0.29	5.53	24.47	0.910	23.45	5.012	1.562	10.03
170.00	1.40	2.94	5.87	0.31	5.25	24.75	0.917	25.34	4.754	1.466	9.89
180.00,	1.20	1.62	3.23	0.33	4.99	25.01	0.924	27.24	4.522	1.379	9.76
190.00	0.89	-1.06	-2.12	0.35	4.75	25.25	0.930	29.15	4.310	1.299	9.62
200.00	1.60	4.06	8.11	0.36	4.53	25.47	0.936	31.06	4.117	1.227	9.50

APPENDIX C

The Smooth Spherical-Earth Model for the Intermediate and Diffraction Regions*

This program, written in FORTRAN IV, computes propagation over a smooth spherical earth using the analysis described in Fock (B1965), Chapter 12. The program assumes that atmospheric refraction is described by the earth-radius factor K and that targets and antennas are at low altitudes. For *horizontally* polarized waves there are no restrictions on the wave frequency, but for vertical polarization the results are not to be trusted below critical frequencies that depend on the conductivity of the ground. These critical frequencies are roughly 300 KHz for dry ground, 3 MHz for moist ground, and 150 MHz for sea water. The program sums a series involving Airy functions of complex argument, and these functions are calculated by integrating the appropriate differential equation. The calculation breaks down, and the series converges excessively slowly, or perhaps diverges, when the line of sight between antenna and target clears the earth's bulge by too much. Although the integration step-size DELTA is set equal to 0.02 in this program, we have found that this step-size is too large in some cases. For computations above the maximum of the lowest lobe we recommend that a step-size of 0.002 be used. This will increase the computation time on a large computer such as the IBM 470 from about 1 second to 10 seconds.

Input

Target range (km) S

Transmitting antenna height H1

Target height (m) H2

Wavelength (m) WAVE

Output

Transmission loss (dB) relative to free space: EXCESS LOSS + XXX.X DB (A loss is indicated by a negative sign)

If the series diverges or converges excessively slowly: SERIES DOES NOT CONVERGE

PROGRAM LISTING: SPHERICAL-EARTH MODEL, INTERMEDIATE AND DIFFRACTION REGIONS

PROGRAMMED BY IRA GILBERT & ANDREA CURTIS

```
      REAL*8 ARRAY(5), Y(2), LOSS, FREE, NET, D, D2, D3, S, TAU, TEMP
      COMPLEX*16 PHI,T,TEE,SUM,W(2),P
      PI=3.1415926536
      DELTA=0.02
C     INTEGRATION STEP FOR GENERATING AIRY FUNCTION
      EPS=.0001
C     ACCURACY CRITERION FOR TERMINATING SERIES
      MAX=101
C     MAXIMUM NUMBER OF TERMS IN SERIES
1     WRITE(6,100)
100   FORMAT(' ENTER RANGE(KM),H1(M),H2(M),WAVELENGTH(M),EARTH-RADIUS
     *FACTOR')
      ARRAY(1)=0.
      CALL READV(ARRAY)
      NUMBER=ARRAY(1)
      IF(NUMBER .EQ. 0) CALL EXIT
C     READV IS A FORMAT-FREE INPUT ROUTINE, ARRAY THE INPUT DATA.
      A=6378388.
C     RADIUS OF EARTH
      S=ARRAY(1)*1000.
      H1=ARRAY(2)
      H2=ARRAY(3)
      WAVE=2.*PI/ARRAY(4)
      IF(ARRAY(5) .NE. 0.) A=A*ARRAY(5)
C     S IS THE DISTANCE BETWEEN ANTENNAS, H1 AND H2 THEIR HEIGHTS,
C     AND WAVE THE WAVENUMBER OF THE SIGNAL ALL IN METERS.
      EM=(WAVE*A/2.)**(1./3.)
      X=S*EM/A
      Y(1)=WAVE*H1/EM
      Y(2)=WAVE*H2/EM
C     X, Y(1), AND Y(2), ARE DIMENSIONLESS VARIABLES RELATED TO
C     ANTENNA SEPARATION AND HEIGHT.
      SUM=0.
      DO 40 I=1,MAX
```

```
C      COMPUTING ROOTS OF AIRY FUNCTION
       IF (I.EQ.1) TAU=2.33811
       IF (I.EQ.2) TAU=4.08795
       IF (I.EQ.3) TAU=5.52056
       IF (I.EQ.4) TAU=6.78671
       IF (I.EQ.5) TAU=7.94417
       IF (I.LE.5) GO TO 10
       M=4*I-1
       M2=M*M
       Z=M*(PI/4+(0.0884194-(0.08328-0.4065/M2)/M2)/M2)
       TAU=(1.5*Z)**(2./3.)
10     T=DCMPLX(.5*TAU,.8660254*TAU)
C      T IS THE I'TH COMPLEX ROOT OF THE AIRY FUNCTION
       PHI=CDEXP(X*(0.,1.)*T)
C      GENERATING THE AIRY FUNCTIONS BY NUMERICALLY INTEGRATING
C      THE DEFINING DIFFERENTIAL EQUATION
       DO 30 K=1,2
       N=Y(K)/DELTA+1
       D=-Y(K)/N
       D2=D*D/2
       D3=D2*D/3
       TEE=T
       W(K)=0.
       P=1.0
       DO 30 J=1,N
       W(K)=W(K)*(1.+D2*TEE+D3)+P*(D+D3*TEE)
       P=P*(1.+D2*TEE+D3*2.)+W(K)*(D*TEE+D2+D3*(TEE*TEE))
30     TEE=TEE+D
C      WRITE(6,13)W(1),W(2)
13     FORMAT(2X,'W(1)=',E10.2,'  W(2)=',E10.2)
       SUM=SUM+PHI*W(1)*W(2)
       IF (I.GT.5) GO TO 35
C      WE REQUIRE THE SERIES TO CONTAIN AT LEAST 5 TERMS
       GO TO 40
```

```
35      TEMP=LOSS
        LOSS=4.*PI*X*(CDABS(SUM))**2
        NET = CDABS(SUM)
C       WRITE(6,12)X,NET,LOSS
12      FORMAT(2X,'X=',E10.2,'  CDABS(SUM)=',E10.2,'  LOSS=',E10.2)
        ERROR=DABS(TEMP-LOSS)/LOSS
        IF (ERROR.LT.EPS) GO TO 50
C       TEMP IS THE LOSS CALCULATED WITH I-1 TERMS, LOSS THE
C       LATEST VALUE, AND ERROR THE FRACTIONAL CHANGE.
C       WHEN THE CHANGE IS LESS THAN EPS, THE SERIES IS TERMINATED.
40      CONTINUE
50      NET=-10.*DLOG10(LOSS)
        LOSS=NET
C       ROUNDING LOSS TO NEAREST .1 DB
        IF (I.LT.MAX) WRITE (6,120) LOSS
120     FORMAT(' EXCESS LOSS =', F5.1,' DB')
        IF (I.EQ.MAX) WRITE (6, 130)
130     FORMAT(' SERIES DOES NOT CONVERGE')
        GO TO 1
        END
```

SAMPLE OUTPUT: SPHERICAL-EARTH MODEL, INTERMEDIATE AND DIFFRACTION REGIONS

```
LOAD SPHERE ( CLEAR START
EXECUTION BEGINS...
ENTER RANGE(KM),H1(M),H2(M),WAVELANGTH(M),EARTH-RADIUS FACTOR
6 8 13 .1 1.333
EXCESS LOSS = -4.8 DB
ENTER RANGE(KM),H1(M),H2(M),WAVELANGTH(M),EARTH-RADIUS FACTOR
5 13 13 .1 1.333
EXCESS LOSS = -4.6 DB
ENTER RANGE(KM),H1(M),H2(M),WAVELANGTH(M),EARTH-RADIUS FACTOR
9 26 26 .1 1.333
SERIES DOES NOT CONVERGE
ENTER RANGE(KM),H1(M),H2(M),WAVELANGTH(M),EARTH-RADIUS FACTOR
9 10 10 .1 1.333
EXCESS LOSS = -0.3 DB
ENTER RANGE(KM),H1(M),H2(M),WAVELANGTH(M),EARTH-RADIUS FACTOR
0
R;
```

APPENDIX D

Bibliographical Index: Radar Propagation At Low Altitudes

This bibliography covers the subject of radio and microwave propagation, VHF through X-band, near the surface of the earth. References are indexed under the headings of *Books*, *Journal Articles*, and *Technical Reports* and cross-referenced in a Bibliographical Index. Although many of the references cited here deal with one-way propagation, the results are directly applicable to the radar (two-way) case as explained in Chapter 2. We assembled this bibliography as we surveyed the literature on the effects of propagation on ground-based radar performance against low-flying aircraft. All listed references were examined and found relevant to this general subject: the references judged most useful are indicated by bold face type. In striving for completeness we have included some references of marginal value, but all are worthy of examination if their title or their listing in the Bibliographical Index suggests that they may be of interest. Refererences in the text of this report and in the Bibliographical Index are labeled by the last names of authors and the year of publication. The author listing contains the complete reference information. Major headings in the Bibliographical Index are listed alphabetically below.

1. Diffraction Effects
2. Frequency Bands
3. Height-Gain Measurements
4. Microwave Links
5. Mobile Communication
6. Models for Loss Computation
7. Ocean, Propagation Over
8. Reflection Effects
9. Refraction Effects and Rain Attenuation
10. Review Articles and Reports
11. Spherical Earth, Propagation Over
12. Tracking Accuracy, Propagation Effects on

1. **Diffraction effects**

1.1 cylinders, calculation of diffraction by
Bachynski (J1960)
Capon (R1976)

Dougherty and Maloney (J1964)
Millington (J1960)
Pryce (J1953)
Rice (J1954)
Wait and Conda (J1959)

1.2 cylinder, measurement of diffraction by
Bachynski (J1960)

1.3 experimental studies of
Bullington (J1950b)
Carlson & Waterman (J1966)
Crysdale (J1958)
Day and Trolese (J1950)
Delaney and Meeks (J1979)
Dougherty and Maloney (J1964)
Epstein and Peterson (J1953)
Kirby, Dougherty, and McQuate (J1955)
Lagrone (J1977)
Lagrone, Martin, and Chapman (J1963)
Lelliott and Thurlow (J1965)
Littlewood (J1967)
McPetrie and Ford (J1964)
Neugebauer and Bachynski (J1958)
Neugebauer and Bachynski (J1960)
Oxehufwud (J1959)
Reudink and Wazowicz (J1973)
Rocco and Smith (J1949)
Schlussler (R1973)
Selvidge (J1941)

1.4 knife-edge
Anderson, Trolese, and Weisbrod (J1960)
Bachynski and Kinsmill (J1962)
Kouyoumjian and Pathak (J1974)
Kouyoumjian and Pathak (J1977)
Meeks (J1982)
Ruze, Sheftman and Cahlander (J1966)
Wait and Spies (J1968)

1.5 model measurements
Bachynski and Kinsmill (J1962)

Bachynski (J1963)
Carlson (J1973)
Dougherty (J1970b)
Hacking (J1968)
Hacking (J1970)

1.6 mountains, diffraction over
Carlson and Waterman (J1966)
Carlson (J1973)
Crysdale (J1955)
Dickson *et al.* (J1953)
Kirby, Dougherty, and McQuate (J1955)
Muromoto, Mushiake, and Adachi (J1969)
Neugebauer and Bachynski (J1958)
Neugebauer and Bachynski (J1960)
Shkarofsky, Neugebauer, and Bachynski (J1958)
Swenson (J1956)

1.7 multiple obstacles
De Assis (J1971)
Deygout (J1966)
Haakinson, Violette, and Hufford (R1980)
Kouyoumjian and Pathak (J1974)
Kouyoumjian and Pathak (J1977)
Legg (J1965)
Lopez (J1970)
Millington, Hewitt, and Immirzi (J1962)
Rahmat-Samii and Mittra (J1977)
Wait (J1968)

1.8 theoretical analysis
Dougherty (J1970a)
Dougherty (J1970b)
Furutsu (J1966)
Furutsu and Wilkerson (J1970)
Kouyoumjian and Pathak (J1974)
Kouyoumjian and Pathak (J1977)
Muromoto, Mushiake, and Adachi (J1969)
Ott (J1971)
Pathak, Burnside, and Marhefka (J1980)
Vogler (J1964)
Wait (J1968)

2. Frequency bands

2.1 VHF (30-300 MHz) loss calculations

Adams (J1978)
Barrick (J1971)
Bullington (J1950b)
Crysdale (J1955)
Dickson *et al.* (J1953)
Fink (B1957)
Hortenbach (J1970)
LaGrone (J1960)
LeBay (R1977)
McMahon (J1974)
Ott, Vogler, and Hufford (J1979)
Schelling, Burrows, and Ferrell (J1933)
Swenson (J1956)

2.2 VHF (30-300 MHz) measurements

Brown, Epstein, and Peterson (J1948)
Day and Trolese (J1950)
De Assis (J1971)
Egli (J1957)
Josephson and Blomquist (J1958)
Kirby, Dougherty, and McQuate (J1955)
LaGrone (J1977)
LaGrone, Martin, and Chapman (J1963)
Longley and Hufford (R1975)
Palmer (J1980)
Saxton (J1950)
Saxton and Lane (J1955)
Selvidge (J1941)

2.3 UHF (300-1000 MHz) loss calculations

Adams (J1978)
Bullington (J1950b)
De Assis (J1971)
Fink (B1957)
Hortenbach (J1970)
LaGrone (J1960)
LeBay (R1977)
McMahon (J1974)
Ott, Vogler, and Hufford (J1979)

2.4 UHF (300-1000 MHz) measurements
- Black and Reudink (J1972)
- Brown, Epstein, and Peterson (J1948)
- **Day and Trolese** (J1950)
- **Epstein and Peterson** (J1953)
- Hacking (J1968)
- Hacking (J1970)
- Head (J1960)
- Kirby, Dougherty, and McQuate (J1955)
- Jakes and Reudink (J1967)
- LaGrone (J1977)
- LaGrone, Martin, and Chapman (J1963)
- Longley and Hufford (R1975)
- Palmer (J1980)
- Reudink and Wazowicz (J1973)
- Saxton and Lane (J1955)
- Schlussler (R1973)
- Young (J1952)

2.5 L-Band (1-2 GHz) loss calculations
- Adams (J1978)
- **Meeks** (J1982)

2.6 L-Band (1-2 GHz) measurements
- **Day and Trolese** (J1950)
- LaGrone (J1977)
- Littlewood (J1967)
- **Meeks** (J1982)
- McGavin and Maloney (J1959)
- Saxton and Lane (J1955)
- Straiton (J1952)
- Sun (R1979)

2.7 S-Band (2 -4GHz) measurements
- Carlson and Waterman (J1966)
- **Day and Trolese** (J1950)
- Durkee (J1948)
- LaGrone (J1977)
- LaGrone and Chapman (J1961)
- Littlewood (J1967)
- McPetrie and Ford (J1964)

O'Dowd, Dyer, and Tuley (R1978)
Oxehufwud (J1959)
Sherwood and Ginzton (J1955)
Straiton (J1952)
Young (J1952)

2.8 C-Band (4 - 8 GHz) loss calculations
Bullington (J1950a)
Crane (R1977)
Linlor (J1980)

2.9 C-Band (4 - 8 GHz) measurements
Beard (J1961)
Bullington (J1954)
Durkee (J1948)
Hearson (J1967)
Lelliott and Thurlow (J1965)
Littlewood (J1967)
Oxehufwud (J1959)

2.10 X-Band (8 -12.5 GHz) loss calculations
Barton (J1977)
Crane (R1977)
Linlor (J1980)

2.11 X-Band (8 -12.5 GHz) measurements
Barsis, Barghausen, and Kirby (J1963)
Beard (J1961)
Cumming (J1952)
Day and Trolese (J1950)
Durkee (J1948)
Haakinson, Violette, and Hufford (R1980)
Jakes and Reudink (J1967)
LaGrone and Straiton (J1949)
Legg (J1965)
Littlewood (J1967)
Oxehufwud (J1959)
Reudink (J1972)
Reudink and Wazowicz (J1973)
Straiton (J1952)
Tomlinson and Straiton (J1959)

2.12 higher frequencies (>12.5 GHz) loss calculations
Crane (R1977)

2.13 higher frequencies (>12.5 GHz) measurements
Armstrong, Cornwell, and Greene (R1974)
Beard (J1961)
Blue (J1980)
Day and Trolese (J1950)
Durkee (J1948)
Forbes (J1968)
Haakinson, Violette, and Hufford (R1980)
Hayes, Lammers, *et al.* (J1979)
Straiton and Tolbert (J1956)

3. **Height-gain measurements**

Day and Trolese (J1950)
Delaney and Meeks (J1979)
Epstein and Peterson (J1953)
Haakinson, Violette, and Hufford (R1980)
Hamlin and Gordon (J1948)
Hayes and Lammers (J1979)
Hearson (J1967)
LaGrone and Straiton (J1949)
Legg (J1965)
Lelliott and Thurlow (J1965)
Littlewood (J1967)
Meeks (J1982)
Oxehufwud (J1959)
Rocco and Smith (J1949)
Schlussler (R1973)
Straiton and Tolbert (J1956)
Tomlinson and Straiton (J1959)

4. **Microwave links**

Anderson (J1963)
Bachynski (J1959)
Bullington (J1950a)
Bullington (J1954)
Durkee (J1948)

Forbes (J1968)
Gough (J1962)
Hearson (J1967)
Legg (J1965)
Lelliott and Thurlow (J1965)

5. **Mobile communication**

Black and Reudink (J1972)
Edwards and Durkin (J1969)
Egli (J1957)
Jakes and Reudink (J1967)
Okumara *et al.* (J1968)
Reudink (J1972)
Reudink and Wazowicz (J1973)
Young (J1952)

6. **Models for loss computation**

6.1 computer based
Dadson (J1979)
Delaney and Meeks (J1979)
Edwards and Durkin (J1969)
Electromagnetic Compatibility Analysis Center (R1978)
Longley and Rice (R1968)
Palmer (J1979)

6.2 empirical or semi-empirical
Adams (J1978)
Bullington (J1950a)
Bullington (J1950b)
Bullington (J1954)
Bullington (J1957)
Day and Trolese (J1950)
De Assis (J1971)
Fink (B1957)
Longley and Reasoner (R1970)
Longley, Reasoner, and Fuller (R1971)
Longley and Hufford (R1975)
Longley (R1976)
Longley (R1978)

6.3 theoretical

Anderson, Trolese, and Weisbrod (J1960)
Barton (J1974)
Crysdale (J1955)
Dickson *et al.* (J1953)
Domb and Pryce (J1946)
Hortenbach (J1970)
Kalinin (R1958)
Meeks (J1982)
Norton (J1941)
Norton, Rice, and Vogler (J1955)
Ott, Vogler, and Hufford (J1979)
Rice, Longley, Norton, and Barsis (R1967)
Radio Research Laboratories, Tokyo (J1957)
Vogler (J1964)

7. **Ocean, propagation over**

Barrick (J1971)
Beard et al. (J1956)
Beard (J1961)
Beard (J1967)
DeLorenzo and Cassedy (J1966)
Longuet-Higgins (J1960a)
Longuet-Higgins (J1960b)
Klein and Swift (J1977)
O'Dowd, Dyer, and Tuley (R1978)

8. **Reflection effects**

8.1 electrical properties of surface materials

Blue (J1980)
Carlson (J1967)
Cumming (J1952)
Evans (J1965)
Fung and Ulaby (J1978)
Klein and Swift (J1977)
Linlor (J1980)
Njoku and Kong (J1977)
Stogryn (J1971)
Wang (R1979)
Watt and Maxwell (J1960)

8.2 reflection coefficients, measurements of
- **Beard** (J1961)
- **Bullington** (J1954)
- Clarke and Hendry (J1964)

8.3 reflection coefficients, measurements of
- Cornwell and Landcaster (J1979)
- Cumming (J1952)
- **Day and Trolese** (J1950)
- Durkee (J1948)
- **Ford and Oliver** (J1946)
- Hayes and Lammers (J1979)
- Josephson and Blomquist (J1958)
- Lelliott and Thurlow (J1965)
- McGavin and Maloney (J1959)
- **Mrstik and Smith** (J1978)
- Oxehufwud (J1959)
- Peake and Oliver (R1971)
- **Sherwood and Ginzton** (J1955)
- Straiton (J1952)
- Sun (R1979)

8.4 reflection from rough surfaces, theory of
- Barton (R1976)
- Barton (J1979b)
- Beard *et al.* (J1956)
- Beckmann (J1965)
- Beckmann (R1967)
- Beckmann (J1973)
- Boyd and Deavenport (J1973)
- **Bullington** (J1954)
- Capon (R1976)
- Clarke and Hendry (J1964)
- Davies (J1975)
- DeLorenzo and Cassedy (J1966)
- Hufford (J1952)
- Longuet-Higgins (J1960a)
- Longuet-Higgins (J1960b)
- Ott (J1971)
- Ott and Berry (J1970)
- Saxton (J1950)
- **Smith** (1967)

Twersky (J1957)
Wagner (J1967)

8.5 shadowing effects in low-grazing-angle reflection
Barton (R1976)
Beckmann (J1965)
Brockelman and Hagfors (J1966)
Davies (J1975)
Hearson (J1967)
Lynch and Wagner (J1970)
Smith (J1967)
Wagner (J1967)
Wetzel (J1977)

9. Refraction effects & rain attenuation

Bean, Horn, and Ozanich, Jr. (R1960)
Crane (R1977)
Crawford and Jakes (J1952)
DeLange (J1951)
Durkee (J1948)
Hamlin and Gordon (J1948)
Legg (J1965)
McCue (R1978)
Moene (R1966)
Nottarp (J1967)
Rocco and Smith (J1949)
Samson (R1975)
Samson (R1976)
Segal and Barrington (R1977)
Tomlinson and Straiton (J1959)

10. Review articles & reports

Bachynski (J1959)
Barton (J1974)
Barton (J1977)
Blomquist (J1975)
Bullington (J1957)
Egli (J1953)
Epstein and Peterson (J1953)

Forbes (J1968)
Glenn (J1968)
Gough (J1962)
Kalinin (R1958)
LaGrone (J1960)
Longley and Reasoner (R1970)
Longley, Reasoner, and Fuller (R1971)
McGarty (J1976)
Mrstik and Smith (J1978)
Rice, Longley, Norton, and Barsis (R1967)
Saxton and Lane (J1955)
Schelleng, Burrows, and Ferrell (J1933)

11. **Spherical earth, propagation over**

Domb (J1953)
Domb and Pryce (J1946)
Gilbert and Curtis (J1976)
Norton (J1941)
Norton, Rice, and Vogler (J1955)
Pryce (J1953)
Van der Pol and Bremmer (J1937)
Vogler (J1964)
Wait (J1968)
Wait and Spies (J1968)

12. **Tracking accuracy, propagation effects on**

Armstrong, Cornwell, and Greene (R1974)
Barton (R1976)
Barton (J1977)
Barton (J1979a)
Barton (J1979b)
Cornwell and Landcaster (J1979)
Kammerer and Richer (R1964)
McGarty (R1974)
Mrstik and Smith (J1978)
Smith and Mrstik (J1979)
White (J1974)
Zehner and Tuley (J1979)

REFERENCES

BOOKS

Barton, D.K., (B 1975) *Radars, Volume 4: Radar Resolution and Multipath Effects,* Dedham, Mass.: Artech House, Inc.

Beckmann, P., and A. Spizzichino, (B 1963) *The Scattering of Electromagnetic Waves from Rough Surfaces,* Oxford: Pergammon Press, Ltd. (Distributed by MacMillan Co., New York).

Born, M., and E. Wolf, (B 1959) *Principals of Optics,* London: Pergammon Press, Ltd.

Brodhage, H., and W. Hormuth, (B 1977) *Planning and Engineering of Radio Relay Links,* Eighth Edition, London: Heyden and Son Ltd.

Burrows, C.R., and S.S. Attwood, (B 1949) *Radio Wave Propagation,* New York: Academic Press, Inc.

CCIR, 15th Plenary Assembly, (B 1982) International Radio Consultative Committee (CCIR), Vol. V of the Reports and Recommendations, International Telecommunications Union, Geneva.

Fink, D.G., (B 1957) *Television Engineering Handbook,* (Chapter 14, Wave Propagation, Radiation and Absorption) New York: McGraw-Hill, Inc.

Fock, V.A., (B 1965) *Electromagnetic Diffraction and Propagation Problems,* Oxford: Pergammon Press, Ltd.

Kerr, D.E., editor, (B 1951) *The Propagation of Short Radio Waves,* Vol. 13, Radiation Laboratory Series, New York: McGraw-Hill, Inc.

Kong, J.A., (B 1975) *Theory of Electromagnetic Waves,* New York: John Wiley and Sons, Inc.

Langer, R.E., editor, (B 1962) *Electromagnetic Waves,* Proceedings of a Symposium Conducted by the Mathematics Research Center, University of Wisconsin, Madison, 10-12 April 1961, Madison: University of Wisconsin Press.

Livingston, D.C., (B 1970) *The Physics of Microwave Propagation,* Englewood Cliffs, N.J.: Prentice-Hall, Inc.

Meeks, M.L., editor, (B 1976) *Methods of Experimental Physics,* Volume 12B, Radio Telescopes, New York: Academic Press.

Reed, H.R. and C.M. Russell, (B 1953) *Ultra High Frequency Propagation,* New York: John Wiley and Sons, Inc.

JOURNAL ARTICLES

Adams, B.W.P., (J 1978) "An Empirical Routine for Estimating Reflection Loss in Military Radio Paths in the VHF and UHF Bands," in "International Conference on Antennas and Propagation, Part 2, Propagation, IEE Conference Publication No. 169, London.

Anderson, E.W., (J 1963) "The SHF Radio Link Comes of Age," *Point to Point Communication, 7:* 4-15.

Anderson, L.J., L.G. Trolese, and S. Wesibrod, (J 1960) Simplified Method of Computing Knife-Edge Diffraction in the Shadow Region" in *Electromagnetic Wave Propagation,* edited by M. Desirant and J. Michiels, New York: Academic Press, Inc.

Bachynski, M.P., (J 1959) "Microwave Propagation Over Rough Surfaces," *RCA Rev., 20:* 308-335.

———, (J 1960) "Propagation at Oblique Incidence Over Cylindrical Obstacles, " *J. Res. Nat. Bur. Stand., Sect. D, 64D:* 311-315.

———, (J 1963) "Scale-Model Investigations of Electromagnetic Wave Propagation Over Natural Obstacles," *RCA Rev., 24:* 105-144.

Bachynski, M.P., and M.G. Kingsmill (J 1962) "Effect of Obstacle Profile on Knife-Edge Diffraction," *IRE Trans. Antennas Propag. AP-10,* 201-205.

Barrick, D.E., (J 1971) "Theory of HF and UHF Propagation across the Rough Sea," *Radio Sci., 6:* 517-533.

Barsis, A.P., A.F. Barghausen, and R.S. Kirby (J 1963) "Studies of Within-the-Horizon Propagation at 9300 Mc," *IEEE Trans. Antennas Propag., AP-11:* 24-38.

Barton, D.K., (J 1974) "Low-Angle Radar Tracking," *Proc. IEEE, 62:* 687-704.

Barton, D.K., (J 1977) "Radar Multipath Theory and Experimental Data," *IEEE International Conference Radar-77;* London, October 1977.

———, (J 1979a) "Multipath Fluctuation Effects in Track-While-Scan Radar," *IEEE Trans. Aerospace Electr. Sys. AES-15:* 754-764.

———, (J 1979b) "Low-Altitude Tracking Over Rough Surfaces I: Theoretical Predictions," in EASCON '79 Record, IEEE Electronics and Aerospace Systems Convention, Arlington, Virginia, IEEE Publication 79Ch 1476-1 AES.

Beard, C.I., (J 1961) "Coherent and Incoherent Scattering of Microwaves from the Ocean," *IRE Trans., AP-9:* 470-483.

———, (J1967) "Behavior of Non-Rayleigh Statistics of Microwave Forward Scatter from a Random Water Surface," *IEEE Trans. Antennas Propag., AP-15:* 649-657.

Beard, C.I., I. Katz, and L.M. Spetner, (J 1956) "Phenomological Vector Model of Microwave Reflection from the Ocean," *IRE Antennas Propag., AP-4:* 162-167.

Beckmann, P., (J 1965) "Shadowing of Random Rough Surfaces," *IEEE Trans. Antennas Propag., AP-13:* 384-388.

———, (J 1973) "Scattering by Non-Gaussian Surfaces," *IEEE Trans. Antennas Propag., AP-21:* 169-175.

Black, D.M., and D.O. Reudink, (J 1972) "Some Characteristics of Mobil Radio Propagation at 836 MHz in the Philadelphia Area," *IEEE Trans. Vehicular Tech., VT-21:* 45-51.

Blomquist, A., (J 1975) "Seasonal Effects on Ground-Wave Propagation in Cold Regions," *J. Glaciology, 15:* 285-303.

Blue, M.D., (J 1980) "Permittivity of Ice and Water at Millimeter Wavelengths," *J. Geophys. Res., 85:* 1101-1106.

Boyd, M.L., and R.L. Deavenport (J 1973) "Forward and Specular Scattering from a Rough Surface," *J. Acoust. Soc. Am., 53:* 791-801.

Brockelman, R.A., and T. Hagfors (J 1966) "Note on the Effect of Shadowing on the Backscattering of Waves from a Random Rough Surface," *IEEE Trans. Antennas Propag., AP-14:* 621-629.

Brown G.H., J. Epstein and D. W. Peterson (J 1948) "Comparative Propagation Measurements; Television Transmitters at 67.25, 288, 510 and 910 Megacycles," *RCA Rev., IX:* 177-201.

Bullington, K., (J 1950a) "Propagation of UHF and SHF Waves Beyond the Horizon," *Proc. IRE, 38:* 1221-1222.

Bullington, K., (J 1950b) "Radio Propagation Variations at VHF and UHF," *Proc. IRE, 38:* 27-32.

Bullington, K. (J 1954) "Reflection Coefficients of Irregular Terrain," *Proc. IRE, 42:* 1258-62.

———, (J 1957) "Radio Propagation Fundamentals," *Bell Syst. Tech.J., 36:* 593-626.

Carlson, A., (J 1973) "Shadow-Zone Diffraction Patterns for Triangular Obstacles," *IEEE Trans. Antennas Propag., AP-21:* 121-124.

Carlson, A., and A.T. Waterman (J 1966) "Microwave Propagation Over Mountain-Diffraction Paths," *IEEE Trans. Antenna Propag., AP-14:* 489-496.

Clarke R.H. and G.O. Hendry (J 1964) "Prediction and Measurement of the Coherent and Incoherent Power Reflected from a Rough Surface," *IEEE Trans. Antennas Propag., AP-12:* 353-363.

Cornwell, P.E., and J. Landcaster (J 1979) "Low-Altitude Tracking Over Rough Surfaces II: Experimental and Model Comparisons," *IEEE Trans. Aerospace Electr. Sys.,* in EASCON '79 Record, IEEE Record, IEEE Electronics and Aerospace Systems Convention, Arlington, Virginia, IEEE Publication 79CH 1476-1 AES.

Crawford, A.B., and W.C. Jakes (J 1952) "Selection Fading of Microwaves," *Bell Syst. Tech. J., 31:* 68-90.

Crysdale, J.H., (J 1955) "Correspondence: Comments on Dickson *et al,"* (1953) *Proc. IRE, 43:* 627-628.

———, (J 1958) "Comparison of Some Experimental Terrain Diffraction Losses with Predictions Based on Rice's Theory for Diffraction," *IRE Trans. Antenna Propag., AP-6:* 293-295.

Cumming, W.A., (J 1952) "The Dielectric Properties of Ice and Snow at 3.2 Centimeters," *J. Appl. Phys., 23:* 768-773.

Dadson, C.E., (J 1979) "Radio Network and Radio Link Surveys Derived by Computer from a Terrain Data Base," in AGARD Conference Proceedings No. 269, North Atlantic Treaty Organization, Advisory Group for Aerospace Research and Development.

Davies, H.G., (J 1975) "Scattering of Acoustic Point-Source Fields by Random Surfaces," *J. Acoust. Soc. Am., 57:* 1403-1408.

Day, J.P., and L.G. Trolese, (J 1950) "Propagation of Short Radio Waves Over Dessert Terrain," *Proc. IRE, 38:* 165-175.

DeAssis, M.S., (J 1971) "A Simplified Solution to the Problem of Multiple Diffraction over Rounded Obstacles," *IEEE Trans. Antennas Propag., AP-19:* 292-295.

Delaney, J.R., and M.L. Meeks (J 1979) "Prediction of Radar Coverage Against Very Low Altitude Aircraft," in AGARD Conference Proceedings No. 269, North Atlantic Treaty Organization, Advisory Group for Aerospace Research and Development.

DeLange, O.E., (J 1951) "Propagation Studies at Microwave Frequencies by Means of Very Short Pulses," *Bell Sys. Tech. J., 31:* 91-103.

DeLorenzo, J.D., and E.S. Cassedy (J 1966) "A Study of the Mechanism of Sea Surface Scattering," *IEEE Trans. Antennas and Propag., AP-14:* 611-620.

Deygout, J., (J 1966) "Multiple Knife-Edge Diffraction of Microwaves," *IEEE Trans. Antennas Propag., AP-14:* 480-489.

Dickson, F.H., J.J. Egli, J.W. Gerbstreit, and G.S. Wickizer (J 1953) "Large Reductions of VHF Transmission Loss and Fading by the Presence of a Mountain Obstacle in Beyond-Line-of-Sight Paths," *Proc. IRE, 41:* 967-969.

Domb, C., and M.H.L. Pryce (J 1946) "The Calculation of Field Strengths Over a Spherical Earth," *Inst. Electrical Engineers J., 94:* 325-339.

Domb, C., (J 1953) "Tables of Functions Occuring in the Diffraction of Electromagnetic Waves by the Earth," *Adv. Phys., 5:* 96-102.

Dougherty, H.T., (J 1970a) "The Application of Stationary Phase to Radio Propagation for Finite Limits of Integration," *Radio Sci. 3,* 1-6.

———, (J 1970b) "Diffraction by Irregular Apertures," *Radio Sci., 5:* 55-60.

Dougherty, H.T., and L.J. Maloney (J 1964) "Application of Diffractions by Convex Surfaces," *Radio Sci., 68D:* 239-250.

Durkee, A.L., (J 1948) "Results of Microwave Propagation Tests on a 40-Mile Overland Path," *Proc. IRE, 36:* 197-205.

Edwards, R., and J. Durkin (J 1969) "Computer Prediction of Service Areas for VHF Mobile Radio Networks," *Proc. IEE, 116:* 1493-1500.

Egli, J.J., (J 1957) "Radio Propagation Above 40 MC Over Irregular Terrain," *Proc. IRE, 45:* 1383-1391.

Epstein, J., and D.W. Peterson (J 1953) "An Experimental Study of Wave Propagation at 850 MC," *Proc. IRE, 41:* 595-611.

Evans, S., (J 1965) "Dielectric Properties of Ice and Snow — a Review," *J. Glaciology, 5:* 773-792.

Forbes, E.J., (J 1968) "Planning 13-GHz TV Relay Systems," *Trans. IEEE Broadcasting, BC-14:* 19-24.

Ford, L.H. and R. Oliver (J 1946) "An Experimental Investigation of the Reflection and Absorption of Radiation of 9 cm Wavelength," *Proc. Phys. Soc. London, 58:* 265-280.

Fung, A.K., and F.T. Ulaby (J 1978) "A Scatter Model for Leafy Vegetation," *IEEE Trans. Geo-science Electr., GE-16*: 281-286.

Furutsu, K., (J 1966) "A Statistical Theory of Ridge Diffraction," *Radio Sci., 1:* 74-98.

Furutsu, K. and R. E. Wilkerson (J 1970) "Obstacle Gain in Radio-Wave Propagation over Inhomogeneous Earth," *Proc. IEE, 117* 887-893.

Gilbert, I.H., and A.J. Curtis (J 1976) "Compatibility Between Linear-FM Radars: Sphere, Fortran IV," *Microwave Journal,* Microwave Engineers Handbook and Buyers Guide, 8-11.

Glenn, A.B., (J 1968) "Fading from Irregular Surfaces for Line-of-Sight Communications," *IEEE Trans. on Aerospace & Electronics Syst., AES-4:* 149-163.

Gough, M.W., (J 1962) "Propagation Influences in Microwave Link Operation," *J. British IRE, 24:* 53-72.

Hacking, K., (J 1968) "Optical Diffraction Experiments Simulating Propagation Over Hills at UHF," *Inst. Elect. Eng., London, Conf., 48:* 58-65.

Hacking, K., (J 1970) "UHF Propagation Over Rounded Hills," *Proc. IEE, 117:* 449-511.

Hamlin, E.W., and W.E. Gordon (J 1948) "Comparison of Calculated and Measured Phase Difference at 3.2 Centimeters Wavelength," *Proc. IRE, 36:* 1218-23.

Hayes, D.T., U.H. Lammers, R.A. Marr, and J.J. McNally (J 1979) "Millimeter Propagation Measurements Over Snow," in EAS-CON '79 Record, *IEEE Electronics and Aerospace Systems Convention,* Arlington, Virginia, IEEE Publication 79CH 1476-1 AES.

Head, H.T., (J 1960) "The Influcence of Trees on Television Field Strengths at Ultra-High Frequencies," *Proc. IRE, 48:* 1016-1020.

Hearson, L.T., (J 1967) "Unusual Propagation Factors in Point-to-Point Microwave System Performance," *IEEE Trans. Communication Technology, COM-15:* 615-625.

Hortenbach, J., (J 1970) "Multiple Ground Reflection Effects on Fading Behavior of VHF/UHF Satellite Transmissions," *IEEE Trans. Antennas Propag., AP-18:* 836-838.

Hufford, G.A., (J 1952) "An Integral Equation Approach to the Problems of Wave Propagation Over an Irregular Surface," *Quarterly J. Applied Math, 9:* 391-404.

Jakes, W.C., Jr., and D.O. Reudink (J 1967) "Comparison of Mobile Radio Transmission at UHF and X-Band," *IEEE Trans. Vehicular Tech., VT-16:* 10-14.

Josephson, B., and A. Blomquist (J 1958) "The Influence of Moisture in the Ground, Temperature, and Terrain on Ground Wave Propagation in the VHF-Band," *IRE Trans. Antennas Propag., AP-6:* 169-172.

Kirby, R.S., H.T. Dougherty and P.L. McQuate (J 1955) "Obstacle Gain Measurements Over Pikes Peak at 60 to 1,046Mc," *Proc. IRE, 43:* 1467-1472.

Kouyomjian, R.G., and H. Pathak (J 1974) "A Uniform Geometrical Theory of Diffraction for an Edge in a Perfectly Conducting Surface," *IEEE Proc., 62:* 1448-1461.

———, (J 1977) "Comments on a Uniform Geormetrical Theory of Diffraction for an Edge in a Perfectly Conducting Surface," *IEEE Trans. Antennas Propag., AP-25:* 447-451.

Klein, L.A., and C.T. Swift (J 1977) "Improved Model for the Dielectric Constant of Sea Water at Microwave Frequencies," *IEEE Trans. Antennas Propag., AP-25*: 104-111.

LaGrone, A.H., (J 1960) "Forecasting Television Service Fields," *Proc. IRE, 48:* 1009-1016.

———, (J 1977) "Propagation of VHF and UHF Electromagnetic Waves over a Grove of Trees in Leaf," *IEEE Trans. Antennas Propag., AP-25:* 866-869.

LaGrone, A.H., and A.W. Straiton (J 1949) "The Effect of Antenna Size and Height Above Ground on Pointing for Maximum Signal," *Proc. IRE, 37:* 1438-42.

LaGrone, A.H., and C.W. Chapman (J 1961) "Some Propagation Characteristics of High UHF Signals in the Immediate Vicinity of Trees," *IRE Trans. Ant. Prop., AP-9:* 487-491.

LaGrone, A.H., P.E. Martin, and C.W. Chapman,(J 1963) "Height Gain Measurements at VHF & UHF Behind a Grove of Trees," *IEEE Trans. Broadcasting, BC-9:* 37-53.

Legg, A.J., (J 1965) "Propagation Measurements at 11 Gc/s Over a 35 Near-Optical Path Involving Diffraction at Two Obstacles," *Electron. Lett., 1:* 285-286.

Lelliott, S.R., and E.W. Thurlow (J 1965) "Path Testing for Microwave Radio-Relay Links," *Post Office Electrical Engineers Journal, 58:* 26-31.

Linlor, W.I. (J 1980) "Permittivity and Attenuation of Wet Snow Between 4 and 12 GHz," *J. Appl. Phys., 51:* 2811-2816.

Littlewood, C.A., (J 1967) "Microwave Path Testing," *Telephony, 172:* 16-18.

Lonquet-Higgins, M.S., (J 1960a) "Reflection and Refraction at a Random Moving Surface. I. Pattern and Paths of Specular Points," *J. Opt. Soc. Am., 50:* 838-844.

———, (J 1960b) "Reflection and Refraction at a Random Moving Surface. II. Number of Specular Points in a Gaussian Surface," *J. Opt. Soc. Am., 50:* 845-850.

Lopez, A.R., (J 1970) "Ray-Diffraction Method for Handling Complex Incident Fields," *IEEE Trans. Antennas Propag., AP-18:* 821-823.

Lynch, P.J., and R.J. Wagner (J 1970) "Rough Surface Scattering: Shadowing, Multiple Scattering and Energy Conservation," *J. Math. Phys., 11:* 3032-3042.

McGarty, T.P., (J 1976) "Antenna Performance in the Presence of Diffuse Multipath," *IEEE Trans. Aerospace and Electr. Sys., AES-12:* 42-54.

McGavin, R.E, and L.J. Maloney, (J 1959) "Study at 1,046 Megacycles per Second of the Reflection Coefficient of Irregular Terrain at Grazing," *J. Res. Nat. Bur. Stand., Sect., D. Radio Propagation, 63D:* 235-248.

McMahon, J.H., (J 1974) "Interference and Propagation Formulas and Tables Used in the Federal Communications Commission Spectrum Management Task Force Land Mobile Frequency Assignment Model," *IEEE Trans. Vehic Tech., VT-23:* 129-134.

McPetrie, J.S., and L.H. Ford (1964) "Some Experiments on the Propagation Over Land of Radiation of 9.2-Wavelength, Especially on the Effect of Obstacles," *J. IEE, 93:* 531-538.

Meeks, M.L., (J 1982) "A Propagation Experiment Combining Reflection and Diffraction," *IEEE Trans. Antennas Propag., AP-30:* 318-320.

Millington, G., (J 1960) "A Note on Diffraction Round a Sphere or Cylinder," *Marconi Rec., 23:* 170-182.

Millington, G., R. Hewitt, and F.S. Immirzi (J 1962) "Double Knife-Edge Diffraction in Field-Strength Predictions," *Proc. IEE, 109C:* 419-429.

Mrstik, A.V., and P.G. Smith (J 1978) "Multipath Limitations on Low-Angle Radar Tracking," *IEEE Trans. Aerospace Electr. Sys., AES-14:* 85-102.

Muromoto, H., Y. Mushiake and S. Adachi (J 1969) "Intensity and Fluctuation of Diffracted Electric Field Behind Polygonally Approximated Mountain Edge," *Electronics and Communications in Japan, AP-52-B:* 51-59.

Neugebauer, H.E.J., and M.P. Bachynski (J 1958) "Diffraction by Smooth Cylindrical Mountains," *Proc. IRE, 46:* 1619-1627.

———J 1960, "Diffraction by Smooth Conical Obstacles," *J. Res. Nat. Bur. Stand., Sect. D, 64D:* 317-329.

Njoku, E.G., and J.A. Kong (J 1977) "Theory for Passive Microwave Remote Sensing of Near-Surface Soil Moisture," *J. Geophys. Res., 82:* 3108-3118.

Norton, K.A., (J 1941) "The Calculation of Ground-Wave Field Intensity Over a Finitely Conducting Spherical Earth," *Proc. IRE, 29:* 623-639.

Norton, K.A., P.L. Rice, and L.E. Vogler (J 1955) "The Use of Angular Distance in Estimating Transmission Loss and Fading Range for Propagation through a Turburlance Atmosphere over Irregular Terrain," *Proc. IRE, 43:* 1488-1521.

Nottarp, Von K., (J 1967) "Zur Mikrowellenrefraktion uber Schnee- und Sandflachen" (On Microwave Refraction Over Snow and Sand), *Nachrichten aus dem Karten Vermesungswesen, Ser. 1:* 35. 35.

Okumura, Y., T. Kawano, and K. Fukuda (J 1968) "Field Strength and Its Variability in VHF and UHF Land-Mobile Radio Service," *Rev. Electrical Communication Lab., 16:* 825-873.

Ott, R.H., (J 1971) "An Alternative Integral Equation for Propagation Over Irregular Terrain, 2," *Radio Sci., 6:* 429-435.

Ott, R.H., and L.A. Berry, (J 1970) "An Alternative Integral Equation for Propagation Over Irregular Terrain," *Radio Sci., 5:* 767-777.

Ott, R.H., L.E. Vogler, and G.A. Hufford (J 1979) "Ground-Wave Propagation Over Irregular Inhomogeneous Terrain: Comparisons of Calculations and Measurements," *IEEE Trans. Antennas Propag., AP-27:* 284-286.

Oxehufwud, A., (J 1959) "Tests Conducted Over Highly Reflective Terrain at 4,000, 6,000, and 11,000 Megacycles," *Trans. American IEE, 78:* 265-270.

Palmer F.H., (J 1979) "VHF/UHF Path Loss Calculations Using Terrain Profiles Deduced from a Digitial Topographic Data Base," in AGARD Conference Proceedings No. 269 North Atlantic Organization Advisory Group for Aerospace Research and Development, September 1979.

———, (J 1980) "Measurements of VHF/UHF Propagation Characteristics over Arctic Paths," *IEEE Trans. Antennas Propag., AP-28:* 733-743.

Pathak, P.H., W.D. Burnside, R.H. Marhefka (J 1980) "A Uniform GTD Analysis of the Diffraction of Electromagnetic Waves by a Smooth Convex Surface," *IEEE Trans. Antennas Propag., AP-28:* 631-642.

Pryce, M.H.L., (J 1953) "Diffraction of Radio Waves by the Curvature of the Earth," *Advan. Phys., 2:* 67-95.

Rahmat-Samii, Y., and R. Mittra (J 1977) "On the Investigation of Diffracted Fields at the Shadow Boundaries of Staggered Paralled Plates — A Spectral Domain Approach," *Radio Science, 12:* 659-670.

Reudink, D.O. (J 1972) "Comparison of Radio Transmission at X-Band Frequencies in Suburban and Urban Areas," *IEEE Trans. Antennas Propag., AP-20:* 470-473.

Reudink, D.O., and M.F. Wazowicz (J 1973) "Some Propagation Experiments Relating Foliage Loss at X-Band and UHF Frequencies," 1198-1206.

Rice, S.O., (J 1954) "Diffraction of Plane Radio Waves by a Parabolic Cylinder," *Bell Syst. Tech. J., 33:* 417-504.

Rocco, M.D., and J.B. Smith (J 1949) "Diffraction of High-Frequency Radio Waves Around the Earth," *Proc. IRE, 37:* 1195-1203.

Ruze, J., F.I. Sheftman, and D.A. Cahlander (J1966) "Radar Ground-Clutter Shields," *Proc. IEEE, 54:* 1171-1183.

Saxton, J.A., (J 1950) "Reflection Coefficient of Snow and Ice at V.H.F.," *Wireless Engineer., 27:* 17-25.

Saxton, J.A., and J.A. Lane (J 1955) "V.H.F. and U.H.F. Reception," *Wireless World, 61:* 229-232.

Schelleng, J.C., Burrows and E.B. Ferrell (J 1933) "Ultra-Short-Wave Propagation," *Proc. IRE, 21:* 427-463.

Selvidge, H., (J 1941) "Diffraction Measurements at Ultra-High Frequencies," *Proc. IRE, 29:* 10-16.

Sherwood, E.M., and E.L. Ginzton (J 1955) "Reflection Coefficients of Irregular Terrain at 10 cm," *Proc. IRE, 43:* 877-878.

Shkarofsky, I.P., H.E.J. Neugebauer, and M.P. Bachynski (J 1958) "Effects of Mountains with Smooth Crests on Wave Propagation," *IRE Trans. Antennas Propag., AP-6:* 341-348.

Smith, B.G., (J 1967) "Geometrical Shadowing of Random Rough Surfaces," *IEEE Trans. Antennas Propag., AP-15:* 668-671.

Smith, P.G. and A.V. Mrstik (J 1979) "Multipath Tracking Errors in Elevation-Scanning and Monopulse Radars," *IEEE Trans. Aerospace Electr. Sys., AES-15:* 765-776.

Stogryn, A., (J 1971) "Equations for Calculating the Dielectric Constant of Saline Water," *IEEE Trans. Microwave Theory Tech., MT-19:* 733-735.

Straiton, A.W., (J 1952) "Microwave Radio Reflection from Ground and Water Surfaces," *IRE Trans. Antennas Propag. PGAP-4,* 37-45.

Straiton, A.W., and C.W. Tolbert (J 1956) "Measurement and Analysis of Instantaneous Radio Height-Gain Curves at 8.6 Millimeters Over Rough Surfaces," *IRE Trans. Antennas Propag., AP-4:* 346-351.

Stutzman, W.L., F.W. Colliver, and H.S. Crawford (J 1979) "Microwave Transmission Measurements for Estimation of the Weight of Standing Pine Trees," *IEEE Trans. Antennas Propag., AP-27:* 22-26.

Swenson, G.W., (J 1956) "VHF Diffraction by Mountains of the Alaska Range," *Proc. IRE, AP-38:* 1049-1050.

Tomlinson, H.T., and A.W. Straiton (J 1959) "Analysis of 3-cm Radio Height-Gain Curves Taken Over Rough Terrain," *IRE Trans. Antennas Propag., AP-7:* 405-413.

Twersky, V., (J 1957) "On Scattering and Reflection of Electromagnetic Waves by Rough Surfaces," *IRE Trans. Antennas Propag., AP-5:* 81-90.

Van der Pol, D., and H. Bremmer (J 1937) "The Diffraction of Electromagnetic Waves . . . Round a Finitely Conducting Sphere . . . ," *Philos, Mag. 24:* Part 1, 141-176; *24:* Part 2, 825-864.

Vogler, L.E., (J 1964) "Calculation of Ground Wave Attenuation in the Far Diffraction Region," *Radio Sci., D-68:* 819-826.

Wagner, R.J., (J 1967) "Shadowing of Randomly Rough Surfaces," *J. Acoust. Soc. Am., 41:* 138-146.

Wait, J.R., (J 1968) "Diffraction and Scattering of the Electromagnetic Groundwave by Terrain Features," *Radio Sci. 3:* 995-1003.

Wait, J.R., and A.M. Conda (J 1959) "Diffraction of Electromagnetic Waves by Smooth Obstacles for Grazing Angles," *J. Res. Nat. Bur. Stand., 63-D:* 181-197.

Wait, J.R., and K.P. Spies (J 1968) "On the Diffraction by a Knife-Edge Obstacle on a Conducting Earth," *Radio Sci., 3:* 1179-1181.

Watt, A.D., and E.L. Maxwell (J 1960) "Measured Electrical Properties of Snow and Glacial Ice," *J. Res. Nat. Bur. Stand., Sect. D, 64D:* 357-363.

Wetzel, L.B., (J 1977) "A Model for Sea Backscatter Intermittency at Extreme Grazing Angles," *Radio Sci. 12,* 749-756.

White, W.D., (J 1974) "Low-Angle Radar Tracking in the Presence of Multipath," *IEEE Trans. Aerospace Electr. Sys., AES-10:* 835-852.

Young, W.R., (J 1952) "Comparison of Mobile Radio Transmission at 150, 450, 900, and 3700 Mc," *Bell Syst. Tech. J., 31:* 1068-1085.

TECHNICAL REPORTS

Barton, D. K., (R 1976) "Forward Scatter at Low Grazing Angles,"in "DARPA Low-Angle Tracking Symposium," Defense Advanced Research Projects Agency, General Research Corp., Santa Barbara, California, B020494.

Bean, B. R., J. D. Horn, and A. M. Ozanich, Jr. (R 1960) "Climatic Charts and Data of the Radio Refractive Index for the United States and the World, Monograph 22," United States Department of Commerce, National Bureau of Standards, Washington, D.C.

Beckmann, P., (R 1967) "Scattering from Rough Surfaces for Finite Distances Between Transmitter and Receiver," Geo-Astrophysics Laboratory, Boeing Scientific Research Laboratories, Seattle, Washington, AD 666 593.

Capon, J., (R 1976) "Multipath Parameter Computations for the MLS Simulation Computer Program," Final Report FAA-RD-76-55, Massachusetts Institute of Technology, Lincoln Laboratory, Lexington, Massachusetts, AD-A024350.

Carlson, N. L., (R 1967) "Dielectric Constant of Vegetation at 8.5 GHz," Technical Report 1903-5, Ohio State University, Electro-Science Laboratory, Columbus, Ohio.

Crane, R. K., (R 1977) "Microwave Scattering Parameters for New England Rain,"Final Technical Report 426, Massachusetts Institute of Technology, Lincoln Laboratory, Lexington, Massachusetts, AD-647798.

Electromagnetic Compatability Analysis Center, (R 1978) "ECAC Capabilities for Terrain-Radio Coverage Analysis," (no report number given) Electromagnetic Compatability Research Center, Annapolis, Maryland.

Haakinson, E. J., E. J. Violette, and G. A. Hufford (R 1980) "Propagation Effects on an Intervisibility Measurement System," NTIA-Report-80-35, Department of Commerce, Institute for Telecommunications Sciences, Boulder, Colorado (February 1980)

Kalinin, A. N., (R 1958) "Approximate Methods of Calculating the Field Strength of Ultra Short Waves Taking Into Account the Influence of Local Terrain," Translation 6005, National Bureau of Standards, Boulder, Colorado (September 1958)

Kammerer, J. E., and K. A. Richer (R 1964) "4.4 mm Near-Earth Antenna Multipath Pointing Errors," Memorandum Report No. 1559, Ballistic Research Laboratories, Aberdeen Proving Ground, Maryland, AD 443 211.

LeBay, P. M., III, (R 1977) "Low Frequency Radar Systems Should Replace High Frequency Radar Systems on the Battlefield to Optimize the Army's Ground Surveillance Radar Capability," Master's Degree Thesis, U.S. Army Command and General Staff College, Fort Leavenworth, Kansas, AD-A043739.

Longley, A. G., R1976, "Location Variability of Transmission Loss — Land Mobile and Broadcast Systems," OT-Report 76-87, Institute for Telecommunications Sciences, U.S. Department of Commerce, Boulder, Colorado, PB 254 472.

———, (R 1978)"Radio Propagation in Urban Areas," Report 78-144, U.S. Department of Commerce, Office of Telecommunications, Boulder, Colorado, PB 281 932.

Longley, A. G., and G. A. Hufford (R 1975) "Sensor Path Loss Measurements Analysis and Comparison with Propagation Models," OT Report 75-74, Institute for Telecommunications Sciences, Office of Telecommunications, U.S. Department of Commerce, Boulder, Colorado, PB 247 638.

Longley, A. G., and R. K. Reasoner (R 1970) "Comparison of Propagation Measurements with Predicted Values in the 20 to 10,000 MHZ Range," Technical Report ERL148-ITS 97, Institute for Telecommunications Sciences, Boulder, Colorado, AD 703 579.

Longley, A. G., R. K. Reasoner, and V. L. Fuller (R1977) "Measured and Predicted Long-Term Distributions of Tropospheric Transmission Loss," Report OT/TRER 16, Institute for Telecommunications Sciences, Boulder, Colorado, COM 75-112 05/2ST.

Longley, A. G. and P. L. Rice (R 1968) "Prediction of Tropospheric Radio Transmission Loss over Irregular Terrain — a Computer Method, 1968," Department of Commerce, ESSD Research Laboratories Report ERL 79-ITS 67, U.S. Government Printing Office. *Modification* to this computer model published as an Appendix to Department of Commerce, Office of Telecommunications Report 78-144, (April 1978).

McCue, J. J. G., (R 1978) "The Evaporation Duct and Its Implications for Low-Altitude Propagation at Kwajalein," Technical Note 1978-6, Lincoln Laboratory, Massachusetts Institute of Technology, Lexington, Massachusetts (11 May 1979) ADA 057 117.

McGarty, T. P., (R1974) "Models of Multipath Propagation Effects in a Ground-to-Air Surveillance System," Technical Note 1974-7, Lincoln Laboratory, Massachusetts Institute of Technology, Lexington, Massachusetts, AD 777 241.

Moene, A., (R 1966) "The Monthly Frequency of Occurrence of Meterological Conditions Favourable for Duct-Propagation of Radar-Waves in the NW-Europe 1965," Interm Report E-79, Norwegian Defense Research Establishment, Kjeller - Norway.

O'Dowd, W. M., Jr., F. B. Dyer, and M. T. Tuley (R 1978) "Effects of the Ocean Surfaces on the Forward Scattering of Radar Signals at Low Incidence Angles," Final Report No. 1873, Georgia Institute of Technology, Engineering Experiment Station, Atlanta, Georgia.

Peake, W. H., and T. L. Oliver (R 1971) "The Response of Terrestrial Surfaces at Microwave Frequencies," Technical Report AFAL-TR-70-301, Air Force Avionics Laboratory, Air Force Systems Command, Wright-Patterson Air Force Base, Ohio.

Radio Research Laboratories, Tokyo, (R 1957) "Atlas of Radio Propagation Curves for Frequencies Between 30 and 10,000 Mc/s," Radio Research Laboratories, Ministry of Posts and Telecommunication, Tokyo, Japan.

Rice, P. L., A. G. Longley, K. A. Norton, and A. P. Barsis (R 1967) "Transmission Loss Predictions for Tropospheric Communication Circuits," Technical Note 101, Volumes I & II, Institute for Telecommunications Sciences and Aeronomy, Boulder, Colorado, AD 687 820 & AD 687 821.

Samson, C.A., (R1975) "Refractivity Gradients in the Northern Hemisphere," OTR 75-59, U.S. Department of Commerce, Office of Telecommunications, Institute for Telecommunications Sciences, Boulder, Colorado, ADA 009 503.

Samson, C. A., (R 1976) "Refractivity and Rainfall Data for Radio Systems Engineering," Department of Commerce, Office of Telecommunications, Report OTR 76-105, (September 1976).

Schussler, H., (R 1973) "UHF Propagation Path Loss Measurements at Low Grazing Angles," Technical Report ECOM-4106, U.S. Army Electronics Command, Fort Monmouth, New Jersey, AD 910 305.

Schouten, A., (R1971) "Attenuation Effect of Foliage Upon Radar-Frequency Propagation — Results of a Trial at Pershore (UK)," Technical Memorandum STC TM-310, Shape Technical Centre, The Hague, AD 888 431.

Segal, B., and R. E. Barrington (R 1977) "The Radio Climatology of Canada: Tropospheric Refractivity Atlas for Canada," Department of Communications (Canada), Communications Research Centre Report No. 1315-E (December 1977).

Sun, D. F., (R1979) "Experimental Measurements of Ground Reflection Elevation Multipath Characteristics and Its Effects on a Small Aperture Elevation Tracking Radar," Technical Note 1979-21, Massachusetts Institute of Technology, Lincoln Laboratory, Lexington, Massachusetts, AD-A977915.

Wang, J. R., (R 1979) "The Dielectric Properties of Soil-Water Mixtures at Microwave Frequencies," Technical Memorandum 80597, NASA Goddard Space Flight Center, Greenbelt, Maryland.

Zehner, S. P., and M. T. Tuley (R 1979) "Development and Validation of Multipath and Clutter Models for TAC ZINGER in Low Altitude Scenarios," Final Report, Engineering Experiment Station, Georgia Institute of Technology, Atlanta, Georgia.

Subject Index

Author Index